An Introduction to Human Operations Psychotherapy

An Introduction to Human Operations Psychotherapy

Steve Davidson

ROWMAN & LITTLEFIELD
Lanham • Boulder • New York • London

Published by Rowman & Littlefield
A wholly owned subsidiary of The Rowman & Littlefield Publishing Group, Inc.
4501 Forbes Boulevard, Suite 200, Lanham, Maryland 20706
www.rowman.com

Unit A, Whitacre Mews, 26-34 Stannary Street, London SE11 4AB

British Library Cataloguing in Publication Information Available

Library of Congress Cataloging-in-Publication Data Available

Names: Davidson, Steve, 1946-author.
Title: An introduction to human operations psychotherapy / Steve Davidson.
Description: Lanham : Rowman & Littiefield, [2017] | Includes bibliographical references and index.
Identifiers: LCCN 2017009889 | ISBN 9781442256637 (cloth : alk. paper) | ISBN 9781442256651 (pbk.: alk. paper) | ISBN 9781442256644 (electronic)
Subjects: | MESH: Psychotherapy | Systems Theory | Personal Satisfaction | Human Development
Classification: LCC RC480 | NLM WM 420 | DDC 616.89/14-dc23
LC record available at https://lccn.loc.gov/2017009889

Printed in the United States of America

To Downs, Wathen, and Pirot,
for your inspiration,
and
to Rebecca Lovell,
for your infinite encouragement.

Contents

Preface

Once upon a time I wanted to be a philosopher. I wanted to be that person, like Nietzsche, wandering across the hills and looking across the Mediterranean Sea, or like Thales, standing on the Ionian shore—pondering life in a fresh and useful way. Then I wanted to be that person in a small room, perhaps by an open window, catching a breeze and the light of the setting sun—writing something that would startle and inspire someone far away. Someone seeking and searching. Someone like me.

It was those big questions that fascinated me. "Who are we?" "Where are we going?" "How do we get there?" And, perhaps most importantly, "When we get there, will they have good delis?"

But after reading around a bit, it started to look as if psychology might have somewhat more directly relevant answers than philosophy. I can remember picking up a slim, brightly colored volume on psychoanalysis. As I recall, the cover announced boldly that the book was going to take the mind right down to its basics and show the reader how to grab ahold of life so that it would produce the results the reader wanted. Pretty heady stuff. And certainly motivating.

However, also as I recall, the book eventually derailed (from my point of view) into a dry, detailed, academic review of concepts and terminology. Not quite what I was looking for. And that process of partially successful searches went on a long, long time.

Because I never found the exact book I wanted to read, I wrote my own!

This book is an attempt to position clinical and industrial/organizational psychology, as well as their personal change relatives, such as education and training, consulting and coaching, within a practical, positive system of living that appeals both to common sense and to the most sophisticated scientific thinking. This book is an attempt to bring into the personal development world some of the brilliant concepts from the field of *operations*, and to show how they can be used to illuminate and ameliorate some of the perennial questions about how to live well and what to do when living isn't quite working.

The material presented here is a composite of philosophy, history, literature, engineering, entertainment, hard science, psychosocial research, law, neuropsychology, clinical experience in psychotherapy and psychological testing, and personal experience such as travel. It is intended to be descriptive as to what seems to work clinically and in life as a whole, not just now but in the past as well. But it is also intended to be motivating, creating a realistic yet inspiring portrait of a high-functioning type of person—an *ops personality*—accompanied by practical steps that could lead to the realization of that personality.

Many people over the years have criticized psychotherapy, sometimes unfairly, as being inadequately grounded in pragmatism and common experience. It is my hope that this book (which is an extension of a prior monograph—Davidson, 2000) will provide a relatively simple, rational, well-grounded formulation of living, of thinking and feeling, of surviving and thriving that will be plausible and helpful to a wide spectrum of readers.

Then, perhaps, those seekers can sit down at a good deli, eat and drink, listen to the music, and talk, and laugh, and dream—and be happy.

Acknowledgments

The number of people to whom I owe thanks for the realization of this book is no doubt large, but I would like to cite some who have been especially thoughtful and helpful, supportive and instructive: Dr. Richard Sorenson, and the I/O program at CSPP/San Diego for its emphasis on precision, modeling, and criterion-referencing, which form the chief framework for the thinking presented in this book; Dr. Douglas Citro, Dr. Julie Cohen, Dr. Kirsten Fleming, Bruce McAllister, Dr. Nina Rodd, Dr. Ronald Smith, Dr. Jack Wasserman, and especially Douglas Downs for reading and commenting on various portions of the manuscript; to my family for its continuous interest, assistance, and faith in a long and difficult process, including my sister Christine who blazed the writing path. I would also like to recognize the meticulous, thoughtful work of the production editor Elaine McGarraugh. And finally, but not by any means least important, my patient, pleasant, insightful, and professional editor for psychology and social work from Rowman & Littlefield, Molly White.

Part I

THEORETICAL FOUNDATION

Chapter 1

Life as an Operation

Means Linked to Ends

"The game is afoot!" (Smith, 2011, p. 129). So begins a famous Sherlock Holmes mystery adventure. It is perhaps not too far off the mark to suggest that life itself is something of a game, an adventure, and a mystery.

This book can be seen as an investigation into that mystery. It holds, likely, few final answers. But it may offer some interesting and useful suggestions as to the nature of life and a means to mastering life, at least to the point of personal satisfaction. In the spirit of science, this book is presented as a system of hypotheses, which can be tested by readers and interested observers and which can be adjusted according to their findings. It is an introduction to a particular way of looking at life, personality, and personal change—*operations*.

A BRIEF HISTORY OF PSYCHOTHERAPY AND PERSONAL CHANGE

Mythology, philosophy, and even literature can be seen as prescientific attempts to understand reality and to cope with it, to understand the world, to order human life sensibly and sanely, and to help create cultures and societies that are stimulating, beautiful, safe, and successful—*psychotherapy in the largest sense*. For example, the early Greek writer Hesiod (1973; also Leeming, 1990) described the goddess of Justice, Themis. She can be seen as simply an appealing member of a colorful mythological pantheon. But she can also be seen as the memorable personification of a brilliant conceptualization of justice: strong and firm, intelligent, and fair, but also compassionate and nurturing—a shrewd blend of paradoxical characteristics (punishment and reinforcement), an attempted upper-level solution to the perennial problem of misbehavior, disorder, and danger. "Themis" amounts to a clever *management principle*—firm but compassionate justice—that can be applied to the community, the organization, the family as well as to the self, to the mind.

Cicero, a Roman statesman, more directly addressed the pursuit of happiness in his work *On the Good Life* (1971). For example, Cicero suggested three particular qualities that would contribute to successful living: (1) developing an accurate and causal view of nature—good reality contact, (2) learning to rationally manage emotions—self-control, and (3) learning to thoughtfully get along with other people—considerateness.

The statements attributed to the ancient sage Confucius (n.d.) are a cornucopia of sensible suggestions as to how humans can live capably and well, such as: (1) nurturing a deep love of studying; (2) being earnest, sincere, calm, and careful in all work; (3) bravely facing difficulties; and (4) continuously refining talents and relationships to live a better and better life.

Even literature can be seen as an entertaining attempt to impart practical wisdom and thus can be seen as conveying psychosocial messages to its audience. For example, the illustrious *Iliad* (Homer, 2004), viewed from one angle, is merely an exciting telling of the Trojan War. But from a more therapeutic angle, the *Iliad* proposes numerous psychosocial guidelines, such as: (1) among the highest virtues are courage, integrity to principles, and sacrifice on behalf of others; (2) grace, fairness, and intelligence can and should be generously admired, even in opponents; (3) persistence and shrewdness are required for success in resolving conflicts; and finally, (4) if something looks too good to be true, it may be; therefore, it is wise to be cautious in evaluating options (like the Trojan Horse).

These examples of essentially psychotherapeutic messages conveyed in myths, philosophies, and literary stories could be multiplied almost infinitely. And, indeed, many of them have been loved and admired for millennia and have functioned as cultural touchstones for entire civilizations. But, as in so many areas, a scientific approach appears to offer, ultimately, the clearest route to mastering a safe and satisfying life.

Science had a long and tangled gestation. The scientific method is one of those developments that looks obvious in hindsight, but was unbelievably difficult to articulate. Initially in China it was technology—aimed at immediately beneficial tools, such as the compass—that tended to occupy state-supported thinkers, rather than an explicit development of theory (Shaughnessy, 2000). In Greece almost the opposite tendency prevailed—an elevation of theory over practice, such as the elaboration of logic and geometry as a means to discovering truth (Curd, 1996; Landels, 1978).

Eventually, through the work of Francis Bacon (1960), Galileo Galilei (1613/1957), and Isaac Newton (Gleick, 2003), among many others, a scientific method was articulated. Subsequently, people in general were astonished at the power of the application of the scientific method to human aspirations and frustrations. Out of nowhere and into their lives, it seemed, roared a stunningly clear explanation of planetary motion, magnetism, and lightning; the invention of the steam engine and locomotives, telescopes, microscopes; and a multitude of other insights and discoveries of the modern world. Naturally, it was not long before scientific thinking was applied to the human mind and its disorders, most prominently, initially, in the person of Sigmund Freud (1930/1961; 1905/1997).

Freud, it seems in hindsight, was a fantastically bold pioneer, in the mid-Victorian era, of trying to apply science to the mind. Unfortunately, despite his many brilliant and useful insights, the science that he was trying to apply was the science of that era—essentially Newtonian mechanics, which described a clockwork universe. The key analogies of the mind, and its disorders, seemed to be an adequately functioning clock, the mechanism of which unfortunately had been blocked, or an adequately formed lock, which had somehow gotten stuck. As Freud said about Dora (1905/1997, p. vii), "The case has opened smoothly to my collection of picklocks." Freud and his followers, such as Carl Jung (1959), eventually developed explanations and treatments of extraordinary and even fascinating complexity, and evolved training programs of great exactitude and length.

It was for those reasons—the extraordinary convolutedness of purported mental mechanisms and the intangibleness of concepts like archetypes, combined with a high degree of single-case observation and speculation—that in the early part of the twentieth century, behaviorists, such as J. B. Watson (1924/1970) of Johns Hopkins and B. F. Skinner (1953) of Harvard, began to focus strictly on observables. The observables, basically, were behavior and environmental stimuli and consequences. Behaviorists conceptually left the contents of the mind a blank, a black box. Their essential premise was that changes in the environment (e.g., reward and punishment) could be instructively correlated with changes in behavior. And, indeed, the extent to which human actions typically correspond to contextual shifts is astonishing (e.g., typically, people go to work primarily to get paid, and if work stops paying, people stop going to work). Behaviorism, and learning theory, which are more amenable to quantification and hypothesis testing, eventually displaced, in large measure, psychodynamic thinking about the human situation in universities and clinics.

However, behaviorism from the beginning relied on animals and children as the foundation of its database (e.g., Pavlov, 1927/1960). Many people have pointed out that animals and children are qualitatively simpler than adult humans (e.g., Piaget & Inhelder, 1969), and thus may form a misleading paradigm. It is clear that people observe, assess, and decide, as well as create and construct, and their thinking, which seems to be at least semi-independent of the environment, directly governs their actions (Luria, 1973). Therefore, it became questionable, in trying to understand human life, to simply skip over what goes on inside the mind. Mental processes, though complex, appear central to human existence and to personal actualization (e.g., Maslow, 1968), and therefore need to be considered. Even the Parisian philosopher Jean-Paul Sartre (1948) was beginning to consider the cognitive superstructure of emotions.

SYSTEMS, COMPUTERS, AND COGNITION

About this same time, in the mid-twentieth century, two theoretical scientific developments, which seem to shed some light on the practical complexities of the mind and of human life, and which came to influence psychotherapy, occurred—systems and computers. Ludwig von Bertalanffy (1968) was a European biologist who was struggling to put together his and other people's observations that life to a great extent coheres in some way, forming identifiable units, with lots of pieces within the unit interacting back and forth productively and at the same time interacting fluidly and adaptively with the environment. In other words, he was articulating *systems theory*. Systems thinking, of course, has grown to encompass multiple aspects of the modern world (e.g., Weinberg, 1980; Yourdon & Constantine, 1979).

It wasn't long at all before systems thinking began to be applied to psychotherapy. Jurgen Ruesch (1951), a psychiatrist, and Gregory Bateson, a highly creative anthropologist, applied the concepts of cybernetics and feedback to human relations and concluded that clear communication loops were the key to healthy functioning. The famed Palo Alto group (Watzlawick, Bavelas, & Jackson, 1967; Watzlawick, Weakland, & Fisch, 1974; Fisch, Weakland, & Segal, 1982) studied language, couples and group interactions, and families from a systems point of view. One of their conclusions was that psychological conflicts get cycled around and around (first-order change—bogus change) and that to resolve conflicts people need to conceptually jump up to a higher level (second-order change—real change) and adopt a new rule for the management of the psychological system.

The second and related development was a long time coming—computers. Figuring things out with machines was an ancient dream. The classical Greeks actually got pretty far with their ingenious Antikythera mechanism (Bertman, 2010)—interlocking, numbered gears designed to predict phases of the moon and the movements of the planets. In the early nineteenth century, Charles Babbage, a young Englishman, famously designed a complex mathematical "difference machine" (Willerding and Hayward, 1968), which never quite worked, but Babbage laid many of the foundations for modern computing. Around the time of World War II, Alan Turing (Moore, 2002) brilliantly realized that a computing machine was essentially a method for making decisions according to an algorithm, a set of if-then rules. In the mid-twentieth century, Ross Ashby (Singh, 1966) proposed the important principle that to be stable, a computing system needed to have multiple governing levels (a hierarchy), so that upper levels could assess and control more immediately reactive lower levels.

Much of this mechanical computational work, of course, carried a leitmotif of trying to understand human thinking; the machines were viewed as clumsy but plausible analogies to the human mind. Mechanical computation was rapidly making enormous progress, so thinking was beginning to take center stage in a lot of theorists' minds. Thus, all this began to mount up into a Kuhnian revolution in psychology (Kuhn, 1962)—too many anomalies accumulated to sustain a radical behaviorist's conception of human life. Thinking—*cognition*—was clearly important in human existence, and the computer researchers were beginning to identify the processes that underlay decision-making and targeted action concretely, dramatically, and convincingly.

The cognitive revolution in psychology in the mid-twentieth century was launched to a marked degree by Jerome Bruner at Harvard (Bruner, Goodnow, & Austin, 1956); by Jean Piaget in Geneva, Switzerland (Piaget & Inhelder, 1969); and by Ulric Neisser at Cornell (1976). If psychodynamic thinking canonized emotions, and if behaviorist thinking canonized behavior, then cognitive psychology canonized thinking itself.

At the University of Pennsylvania in the late 1950s the psychiatrist Aaron Beck (Beck & Young, 1985) was sagely and fortuitously applying cognitive notions to psychotherapy, a program that has been prominently extended in California by the psychologists Dennis Greenberger and Christine Padesky (2016) and that at this point has become a really remarkably successful approach worldwide to managing life difficulties. Cognitive therapy revolves around upsetting beliefs that turn out, on examining the evidence, to be false, or at least doubtful. "Everyone should always like me" would be an example of a depressogenic self-statement that is highly implausible. As clients begin to see the implausibility of depressogenic self-statements, they come into better contact with reality, and depression begins to lift.

ENTER OPERATIONS

Most theories of psychotherapy (Gilliland, 1984) traditionally have been centered on *errors* of functioning (terms such as *psychopathology, pathogenesis, symptom,* and *diagnosis* indicate the framing of the issue). For example, a premise of psychodynamic theory seems to be that the mind is reasonably well constructed and functional other than, primarily, one or more defended, blocked insights, such as a negative transference. Behavioral theory often seems to have as a premise that the black box of the mind is adequate (something like a typical laboratory pigeon), and any behavioral difficulties are due to a misdesign of cues or consequences. And, similarly, a premise of cognitive theory seems to be that the mind is reasonably well constructed and functional other than the acquisition of a number of false upsetting beliefs. These, of course, are broad generalizations about these often complex and constantly evolving psychotherapy theories. But it may not be incorrect to propose that psychotherapy theories tend to presuppose a functioning mind and life marred by identifiable disturbances that become the target of interventions. The assumption seems to be that once the proximate disturbances are corrected (e.g., negative transference, insufficient reinforcers, or false beliefs), the individual will be able to love and work at least adequately. Thus, many of these theories can be seen as *error referenced* rather than *function referenced.*

A radical alternative and perhaps not completely implausible assumption, of course, based partly on nature-nurture research (Polderman et al., 2015) would be that a functional mind has to be partly *constructed* by individuals, families, and societies, building on a biological platform, and that problems in living are largely symptomatic of *lacunae* in the construction process. Perhaps a useful conceptualization here is that the personality may be *co-constructed* in interaction between the individual and the family and then between the individual and society, resulting in an adaptive but unique individual. "Problems in living" would thus become "problems in learning" and then problems in applying that learning to goal-oriented actions.

In any case, it is difficult to find a holistic theory of psychotherapy based on an essentially functional and simple yet comprehensive schema, a basic "how-to" of intelligent living, in which the primary emphasis is on living successfully and the secondary emphasis is on errors that might occur in that process. *The field of operations may provide the basis for such a theory of psychotherapy*; that is the premise of this book.

Operations Per Se

The concept of operations goes way back. The Scottish Enlightenment philosopher David Hume outlined, though in a somewhat scattered fashion, the "operations of the mind" (1748, p. 313). His basic proposal was that people are alert and reasonable, have sensible intentions, and make corresponding plans; then, they exercise their will to take actions necessary to realize their plans.

B. F. Skinner articulated the technology of "operant conditioning" and popularized the idea that organisms carry out effective operations within target contexts. Specifically, he proposed that "the behavior *operates* upon the environment to generate consequences" (1953, p. 65).

Operations has become a huge, mathematically sophisticated, well-articulated field with tremendous impact on a wide variety of endeavors from government and international relations to education and business (Kamauff, 2010; Tomes & Hayes, 1993). The two traditional branches of operations are operations research (OR) and operations management (OM). OR is the more pristine, purely theoretical attempt to identify and understand the best way to accomplish goals. OM is the more hands-on, practical side of getting systems to accomplish selected goals. Operations has been considered to be the premier approach for figuring how to reach goals (OR) and then reaching them (OM). OR and OM *together* are usually referred to as the field of *operations*. Operations can be seen as a treasure chest of practical life-management concepts (briefly reviewed in the following chapter).

Operations theory can be considered a part of systems theory. Systems can be intensely confusing, as they are by definition complex and interactive, with frequent reciprocal effects and tangled communication loops, and with no obvious starting point or ending point.

However, surprisingly, often a system can be pragmatically reduced to a mostly linear model of: Need – Plan – Action – Objective – Feedback. This is similar to the brilliant TOTE (Test – Operate – Test – Exit) model of Miller, Galanter, and Pribram (1960). In other words, often an extensive, kaleidoscoping system (such as, perhaps, a human mind) can be meaningfully distilled down to a core operation, down to rational purposes and corresponding lines of action connecting ideas to results.

A heuristic definition of *operation* is: *an adaptive plan intended to meet needs and desires, transformed into a series of actions aimed at a relevant objective*. An operation thus is fundamentally a scientific process—hypotheses are developed, tested, and adjusted until some semblance of truth is established (epistemology); that notion of truth, that formulation of reality (ontology), then becomes the basis for planning and goal-seeking, within a larger frame of values (axiology).

Here, then, is a basic, imagined example of an operation. A person intuits a need, such as water—thirst. The person hypothesizes what would satisfy the need, such as the water in a nearby stream—that becomes the objective. The person plans a route to the stream, and follows the plan. If the first route doesn't work out, the plan is adjusted, and a secondary route selected and followed, until the stream is reached. Then the water can be drunk, the thirst slaked, and survival (for the moment) achieved.

This example of goal-seeking could be multiplied nearly infinitely. Humans are almost continuously assessing needs, and positing and then pursuing corresponding objectives—doing chores, making friends,

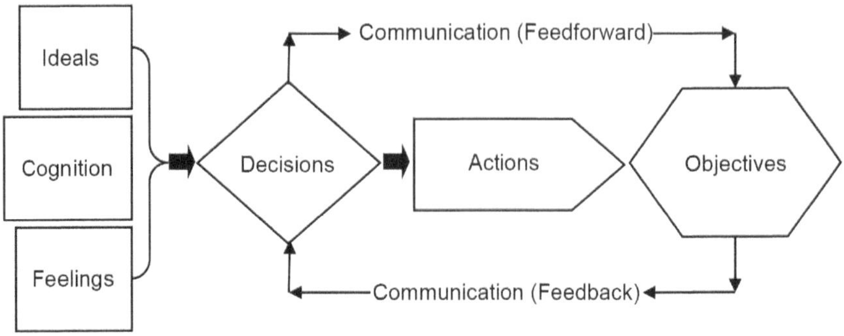

Figure 1.1 Process Model of the Human Mind.

completing homework, getting to school, choosing a career, choosing a training program, getting a job, choosing a mate, establishing a home, raising children, and so on. Living (human life, as well as, perhaps, animal life, and even the functioning of robots) can be seen as a coordinated multiplicity of means aimed at ends. Humans then can be seen as, essentially, *goal-seeking systems*, and the arc of life can be seen as an arc of parallel operations. This conception of a *strategic geometry of living*, that is—*purposeful paths through a life space*—was termed *hodological space* by the innovative MIT social psychologist Kurt Lewin (1951).

If this is so, then the way to succeed in life, in general, is to become good at carrying out operations. Successful living can be conceptualized as flowing from *operational competence*. Then, problems in living would amount to *difficulties in carrying out operations*.

Minds as Meta-operations

If *living* can be conceptualized as a series of operations that satisfy survival needs, then what are *minds*, and why do people have them? As put by Noam Chomsky (2016, 5) "I take the mind here to be the brain viewed at a certain level of abstraction." Minds can be seen as *symbolic systems* that both mirror and superintend physical operations, and have a similar essential structure of means linked to ends—information linked to actions (e.g., Warring, 1984). This is sometimes described as a *mapping process* (Casti, 1989; Damasio, 1994). Minds can be seen as *meta-operations* (Suzuki & Katayama, 1991)—*operations that manage operations*. These informational meta-operations can be seen as essentially scientific, and thus *the mind can be seen as an essentially scientific process*. Figure 1.1 is a heuristic diagram of an ops view of the mind.

In subsequent chapters, figure 1.1 will be elaborated and its meaning and organization clarified, and numerous suggestions will be given as to how to apply this operational model of the mind to living well, and how to deal with difficulties that arise in the course of seeking success in living and loving, relationships and prosperity.

OPERATIONS AS A BROAD PARADIGM

Literature

It is amazing how many stories have an operation as their theme. The concept of *story* itself, including a play, usually has an operational arc—beginning, middle, and end (Taylor, 2002). A quest, such as for the Golden Fleece (Apollonius, 1959) or the *Quest of the Holy Grail* (1969), is an operation. A pilgrimage is an operation, such as that portrayed in Chaucer's *Canterbury Tales* (1958). Exploration and travel themselves are operations, as in the *Odyssey* (Homer, 1963), or the classic poet's journey to northern Japan to see the moon over Matsushima (Basho, 1966). Life is often compared to a journey. The prototypical hero's journey from challenge to conflict through despair to final triumph can be seen as a kind of operation (Campbell, 2004; Seger, 2010). Hollywood script coaches such as Michael Hauge and Christopher Vogler (2003) even talk about an outer journey and a parallel inner journey, perhaps comparable to the physical operation of living superintended by a meta-operation—both are critical in a story, and both are critical to survival.

Tragedies, such as Shakespeare's (1993) *Julius Caesar* or *Romeo and Juliet*, can be seen as *failed operations*. Julius Caesar failed to substantially enhance his level of power without generating fatal opposition, and Romeo and Juliet failed to overcome family conflicts, or to find peace and happiness, or even life, in their romance—in both cases, despite their best efforts.

Humor

Much humor, such as the classic comedies of Charlie Chaplin, Buster Keaton, Laurel and Hardy, and the Marx Brothers, revolves around the inability of ordinary people to adequately carry out the most ordinary kinds of goal-oriented activities, such as walking through a doorway. People losing their identity, their role, or even their gender—such as in Shakespeare's *Twelfth Night* (1993), mistaken identity—is an unsettling source of humor. These absurdities might be called *mistaken operations*.

Mysteries

Mystery stories, such as those of Sherlock Holmes (Smith, 2011) or by Agatha Christie (1934/2011), can be seen largely as *conflicting operations*. The protagonist would like to see justice done, which often means stopping, or even destroying, the antagonist in a prosocial operation. And the antagonist would like to successfully commit the crime and evade the protagonist, in an antisocial operation.

War

War is another form of conflicting operation. For example (to simplify things a bit), one social system, the aggressor, has the intention of taking something from another social system, the victim, or simply punishing the victim in a military operation. The victim then defends itself in a counter-operation. Allies carry out *cooperating operations*.

Sports

Some sports are based on what might be called *competing operations*, such as crossing the finish line first in a race, or being the most efficient in hitting a ball onto a green and then in the hole, in golf. These operations can be seen as analogies to competitive hunting-and-gathering operations—chasing down small, fast, valuable entities and arriving at them more quickly than competitors. Team members carry out cooperating operations.

Other sports are conflicting operations, which might be seen as a kind of symbolic warfare. Both sides are trying to carry out operations (such as make a goal), and both sides are simultaneously trying to keep the other side from being successful. Just so, the two phases, or sides, of these sports are often termed *offense* and *defense*.

Education

Getting an education is an operation—a sequence of learning activities aimed at a diploma. It is a competing operation, in which not only each student strives to reach a standard, but performance is compared with other students. Schools and nations compete, such as in state, national, and international contests and rankings. Conflict is prohibited—actively trying to undercut competitors being seen as at least symbolically déclassé, but, more substantially than that, a net loss to society and to humanity, which needs outstanding educational systems and graduates to survive and thrive.

Business

Business is composed of operations—transforming personnel and materials into profits. Business is normally seen as competing operations, working in parallel to achieve the best outcomes in a high-spirited contest in which society is the final winner (e.g., Adam Smith's discussion of the social benefits of enterprise, 1986). Actually attacking competitors in an offensive, conflicted manner (e.g., industrial espionage and sabotage) is usually seen as inappropriate if not illegal—not in the best interests of society as a whole.

Government and Jurisprudence

Government (e.g., police and military services, diplomatic service, tax collecting, and highway maintenance) can be seen as a system of largely cooperating operations—departments are assigned things they are supposed to be getting done, and they are supposed to work more or less together. That infrastructure (that collection of supporting operations) tends to be the same year in and year out, despite election changes. Politics has become more and more conflicted, versus simply competitive, and perhaps as a result has become less functional. Law can be seen at the judicial level as a cooperating operation (Posner, 1990), in which all the facts are integrated into one decision intended to allow people to operate in a healthy and effective way (i.e., judicial operations support the ideal of justice). Law at the advocacy level is perhaps the prototype conflicting operation—each attorney trying to win at the expense of the other (unless they settle, in a quasi-judicial process).

Relationships

Relationships and romance themselves can be seen as two systems of operations (e.g., both parties have jobs, homes, cars, insurance, bank accounts, and personal goals). In a relationship the two parties agree to ally their operations in some way, to cooperate. In marriage the parties blend their various individual operations into one super-operation, one unified system. Marriage difficulties can be conceptualized as a shift from cooperating operations to competing or even conflicting operations.

The above examples easily could be multiplied many times. This pervasiveness of means-to-ends processes in dramas and comedies, warfare, government and law, sports and business, education and relationships, and almost any human activity that can be named perhaps provides informal but substantial *convergent validity* as to the ubiquitousness of operations in human life. This, then, perhaps, supports the centrality of operational thinking in psychotherapy, the centrality of operations in dealing with problems in living.

SUMMARY

Living can be conceived of as, overall, an operation—a means-to-ends process. Individuals can be conceived of as goal-seeking systems. And the mind can be conceived of as a meta-operation—a symbolic operation that superintends the physical process of applying means to achieve ends, such as meeting needs. The mind can be seen as essentially homologous to other systems that carry out operations—such as economic, family, or political systems, and even biological and robotic systems. All are organizations that carry out operations in, ultimately, a competitive environment of limited resources. Psychotherapy can be seen as a process of sharpening and strengthening the personality, and the goal-seeking that the personality carries out, to increase success and satisfaction in living.

YOU KNOW MY METHODS, WATSON

Over the years, many people have felt that the true attraction of the Sherlock Holmes mysteries is not the stories themselves, but rather the scientific approach to life the stories dramatize, the caring yet objective *methodology* of investigation into puzzles of life and death. Perhaps operations can offer a parallel methodology in the realm of personal growth and change, and in so doing usefully shed light on some of the mysteries of living. Doing that necessitates a clearer, more specific understanding of some of the key concepts of operations.

Chapter 2

Concepts of Operations

From Goal Setting to Independence

Good mysteries often revolve around a clue that is either barely perceptible, such as the footprints outside the window in the Sherlock Holmes story "The Adventure of the Beryl Coronet," or so obvious a missing piece as to be imperceptible, for example, the curiously non-barking dog in the theft of "Silver Blaze" (Smith, 2011). So it is, it would appear, with human existence.

EVERYWHERE, PEOPLE ARE DIFFERENT (OR ARE THEY?)

There's an old mystery about whether people are what they are because of nature, or nurture (Polderman et al., 2015). Are people what they are because of biology, or culture? Correspondingly, a mysterious reality of psychological testing is that it never has 100 percent validity (Kaufman & Lichtenberger, 2006; Nunnally, 1978). A validity of 80 percent would be high. A validity of 60 percent is good and much more common. Why? Why is meticulous, professional testing so partially true? (Or, to put that another way, why are people so *variable*?)

A related mystery that seems to go on and on is the question as to what is the best way to live. It seems like for every country and every era there is a different philosophy (Schaar, 1961). Just when something finally gets worked out, someone else comes along with something to upend it.

For example, Confucius (n.d.) had things pretty well worked out—people should be decent to each other, the master said, and they should internalize principles of good behavior through study and careful observation. Then came along the Legalists to insist that everything should be done according to rigorous debate resulting in regulations and enforcement (Chan, 2014).

The Greek Epicureans' idea was that people should get their lives under control and then eat, drink, and be merry, with special emphasis on the merry (Epicurus, 2012).

But, by contrast, Zen practitioners (Paul, 2000; Suzuki, 1970) made a convincing case that life should be reduced to utter, calm simplicities. The greatest aesthetic satisfaction should come from learning to focus on those simplicities—such as gazing at a subtly dramatic cherry blossom at the end of a long, dark branch; a dark rock among white, raked pebbles; the single sound of a gong echoing through a still morning; or the brisk austerity of a haiku poem.

Thus, there is a tremendous amount of plausible variation in philosophies and styles of living around the world and through time. Often this variation is hard to reliably identify psychometrically because of the difficulty in finding a consistent, core personality template from which to measure variations.

EVERYWHERE, PEOPLE ARE THE SAME

However, just as that truism is established, that everyone everywhere is different and it is hard to find consistencies, it becomes clear that there is *also* an astonishing amount of *commonality* around the world—*universals* (Brown, n.d.). Popular sayings point to this core of similarity, such as: "People are the same all over the world," or "People are people," or "It is just human nature," or "This is the human family." In addition, there is an obvious, and therefore almost invisible (perhaps because of perceptual habituation), similarity across the natural world, and it is clear that the environment tends to shape personality. That is, not only is human biology similar across the world, but the physical human situation is remarkably similar, fundamentally. Specifically, everyone is on the same rotating Earth together; everywhere the sun shines, at least sometimes; almost everyone experiences seasons, rain and heat and cold. Plants and animals, oceans and fish, sky and birds are everywhere similar in form. People mostly *are* water, so they always *need* water, and therefore, they are rarely found too far from rivers, lakes, or oases or at least a working faucet. To reproduce, to protect themselves, and to help each other, people band together, and therefore, everywhere people can be found in groups. And as they navigate through the world, people need rest and sleep. Thus, overall, there seems to be a deep-structure commonality across human groups that is to some extent obscured by the adaptive surface cultural variations.

In a similar vein, a UCLA (University of California, Los Angeles) Sleep Medicine seminar on insomnia treatment pointed out (personal communication, December 7, 2013) that humans are diurnal, *closely linked to the natural world.* Human bodies prefer to get going at dawn (when the light, at the blue end of the spectrum, is activating), and then they start to wind down at sunset (when the light, at the red end of the spectrum, is calming). Then people turn out the lights altogether and go to sleep for about eight hours before dawn. In between dawn and sunset, during the daylight hours, people need to stay active so that they are actually tired at bed time. (The presenters came down hard on late-night television and computer games with all that activating blue light; midnight pizza didn't fare too well, either, with all that caloric energy!)

Thus, there does seem to be a plausible explanation for the mystery as to how it can be that everyone everywhere acts and thinks differently, sometimes radically differently, and yet somehow everyone seems fundamentally the same across the world and through time. As Polderman's research indicated (Polderman et al., 2015), people seem to be about one-half nature and about one-half nurture and learning. Thus, metaphorically, people can be seen as about one-half consistent hardware and about one-half variable software.

But whoever they are, *to survive, people, as living things, have to carry out various operations.* Operations can be seen as a *hidden commonality* among all people, "hidden" because that commonality is so obvious—all around people and in people. People *are* operations—*embodied actions directed toward goals.* Then, different environments lead to different operational styles and traditions.

OPERATIONS INTEGRATES PATTERN AND VARIATION

Reality could be described as *semi-chaotic* (Davidson, 2011). That is, *it is partly structured and partly an unpredictable maelstrom.* Therefore, any system that could adequately master reality has to be prepared (designed, in effect) to deal with both aspects—the predictable and the unpredictable. Not surprisingly, the brain is set up that way, horizontally (left to right) and vertically (top to bottom) (Luria, 1973). To grossly simplify and typify—horizontally, the left side is the structured part (math and language) and the right side is the unstructured part (creativity). Vertically, the upper part (cortex) is the more rational part and the lower part (limbic system) is the more emotional part. Altogether, the brain forms a dynamic system relatively well calibrated (well matched) to both the patterns and the pandemonium of living.

The two chief sectors of operations, OR (operations research) and OM (operations management), can be seen as corresponding to those twin qualities of reality. OR is more theoretical and able to draw broad probabilities out of the maelstroms of life; thus, it is more creative and reflective, more unstructured (something like quantum mechanics). OM is more task oriented; thus, it is more sequential and scheduled, more structured (something like Newtonian mechanics). The two together—operations—form a powerful and adaptive combination for analyzing semi-chaotic situations (that blend of quantum and Newtonian mechanics called *physical reality*) and then for designing activities aimed at reaching objectives within a time frame.

The ops model of the mind (as will be detailed in the next chapter) attempts to capture the twin aspects of reality—order and disorder. The left side of the ops model is Reflection (more creative, more able to deal with and take advantage of disorder, more like OR), and the right side is Execution (more focused, more able to deal with and take advantage of order, more like OM). The rest of this chapter details some of the more important and useful concepts of operations and what might be thought of as a highly specific subfield of operations, project management (Dinsmore, 1990; Kamauff, 2010; Tomes & Hayes, 1993). The idea here is that these concepts can be usefully, even if heuristically, applied to individual, family, and community goal-seeking, that is, to living well, to functioning in an accomplished, confident, healthy manner. These concepts also provide hints as to what to do if things are not going so well, hints as to how to get things back on track.

OPERATIONS CONCEPTS: BROAD

Goals

Goals in the ops approach are defined as *a unified concept linking wants to objectives*. (A quick summation of that definition would be: G = W + O, where *G* means *goal*, *W* means *want*, and *O* means *objective*.) The core of the ops approach is goals. Everything revolves around the decisions as to which needs and desires to fulfill and how to fulfill them. That principle of goal primacy doesn't mean that people can't relax and enjoy themselves, and savor life, love, and laugh—those can be goals, too. It just means that, for people to survive, survival goals have to be set and successfully pursued—food, water, shelter, and so on. In addition, it is surprising to what an extent healthy people pursue multiple goals in their free time (e.g., pursue personal projects and crafts—such as designing and planting a garden, repainting the house, learning to play a musical instrument or to dance, reading all of Agatha Christie's mysteries, or making new friends). Even on vacation, it's usually not long before someone says, "Let's go down to the beach" or "Let's find a great restaurant."

Goals usually exist in hierarchies, something like a classification (an arrangement that also occurs in artificial intelligence; Albus, 1981). Thus, for example, finding the keys fits into the higher goal of driving to town, and driving to town fits into the higher goal of finding a restaurant, and finding a restaurant fits into the higher goal of eating. Upper-tier, governing goals give *meaning* to the subgoals and activities below, and answer the perennial and fundamental question, "Why are we doing this?" Maslow (1970) proposed that there is a hierarchy of system requirements, with more basic requirements being met first and higher, self-actualization tendencies being achieved subsequently.

Goals and goal achievement, to a surprising extent, can be seen as the signatures of people's lives.

Clinical Applications

Traditionally, clients begin by offering a therapist a *presenting problem.* But a *problem* logically exists only in relation to a *goal.* For example, chronic tension is a problem only if the client has the goal of being calmer. Therefore, the primary concern in the ops approach is to identify what goals the client would like to pursue. Almost every clinical problem has a reverse—the goal that defines the problem. The reverse of tension is calmness, the reverse of sadness is cheerfulness, and so on. Almost all presenting problems can be reversed into goals. That reversing transforms psychotherapy overall into a positive process.

"What do you really want?" is the paradigmatic clinical ops question. Not always easy to answer, of course, but that progressive clarification is the beginning of the ops psychotherapeutic process.

Means to Ends

The *end* is the objective, and the *means* is the sequence of actions necessary to reach the end. Thus, *means* is a very general term indicating *process. Ends* is a very general term indicting *purpose* (aim, intent, teleology), either immediate or long term. So, a slightly more formal way of expressing this means-to-ends idea would be *process within purpose.*

The means-to-end thinking is a short-hand, commonsense summary of operational thinking as a whole. *The means-to-end link (M→E) is really the core of an operation* (where *M* means *means*, *E* means *end*, and the arrow indicates a *link*, in which means lead to ends). So, the basic operational question about life is: What must people *do* in order to get what they *want*? Life, over and over, seems to come down to this simple, essential M→E algorithm. This can be seen as a key element in the deep structure of living underlying all the bewildering, mysterious, local, temporary, stylistic variations on the surface. This amounts to an if-then algorithm: If a person wants Y, then that person has to do X, or something equivalent. The usual expression of that, of course, is $y = f(x)$. Ultimately, *operational competence—means-to-ends capability—appears to be what is selected for in the competition of life.*

(That can be summarized by suggesting that the key selection criterion in the competition for survival is: O–C, where *O* means *operational* and C means *competence*.)

Clinical Applications

As painful as it can be for clients to specify what they really want (real needs and deep desires), it can be equally painful, and even a little disheartening, to consider in detail the steps necessitated by a want. A classic example is fitness and weight loss. Most people would like to be thinner and more fit. As everyone knows, the means to that end are all too simple and none too easy—diet and exercise. But weight loss and fitness can be seen as one of the prototypes of all clinical issues—*every goal entails effort*, whether it is transitioning from being tense to being calmer, or transitioning from being sad to being more cheerful, or transitioning from addiction to recovery. Thus, a major part of the ops approach is helping clients cope with the demands of the means-to-end process—the stresses involved in goal achievement.

Operational Focus (Critical Path)

A *critical path* (a concept drawn from project management; Kendrick, 2004) can be interpreted as *the relatively narrow path that efficiently runs from preference toward an objective*. The process of assessing wants, identifying a corresponding objective, and determining a path to that objective sounds pretty straightforward. And, often, it is.

However, it is remarkably easy for people to get derailed, for them to deviate from the critical path. Distractions, fatigue, loss of initial enthusiasm, and discovery of complications in the tasks leading to success, among others, can all encourage deviation from the critical path.

Many people have noted that efficiency seems to fall off, not just directly but exponentially as concentration is broken. Task-oriented thinking, in reality, often involves multiple intents and projects operating on parallel tracks, so to speak—multiple critical paths. Therefore, a key operational skill in maintaining concentration is maintaining separation between critical paths and focusing on just one path at a time until a milestone (a convenient stopping point) has been reached, and then cleanly switching attention and effort to another path. Such a process might be termed *operational clarity*.

Clinical Applications

Clinical situations (presenting problems) can to some extent be described as a general *loss of focus* or at least reduction in focus below the threshold of manageability. Thus, clinical issues as presented are often vague, or conflicting, inexplicable, or transient but recurring—in other words, confusing and frustrating. A major part of the ops approach is to *clarify* the various operations going on in clients' lives and help sharpen clients' awareness of means and ends.

The confusion (the loss of focus) may have been intentional, a means of avoiding painful clarifications and the often demanding focusing of energies (psychodynamically, this would amount to *secondary gain*). Therefore, as before, a major and corresponding part of the ops approach is to *emotionally support* and encourage clients in the often difficult process of building a crisper, more focused operating style.

Pragmatism and Rationality

Operations thinking is *pragmatic*, emphasizing the tangible and the feasible—*facts and logic, practical planning and practical action.* "Where's the beef?" is a popular expression of that idea. Pragmatism has a long and distinguished history, going back at least to William James (1950; 1991) and Charles S. Peirce (1955). Even Confucius (n.d.) and Zen practitioners (e.g., Suzuki, 1970) can be seen as essentially pragmatists (and that practical attitude may partly explain Asia's remarkable success in education, design, and manufacturing).

Rationality, a perhaps broader term and thus more of a conceptual target, has had a long and checkered career in the history of thought, from its canonization by the classical Greeks and the Enlightenment (Hampson, 1968; Outram, 2006) to its virtual trashing by the Romantics (Blanning, 2011). In the ops approach, *rationality* is defined as *a mode of living and thinking in which wants are balanced against efforts, risk is balanced against wisdom, and both are well grounded in empiricism and caring relationships.*

Clinical Applications

Again, many clinical issues, perhaps because of emotional distress leading to cognitive confusion, drift up into the mist of vague speculation, haunting fears, and intractable wish. It is part of the ops process to bring some of these fears and hopes down to the pragmatic ground, where they can be recognized, analyzed, and acted upon and to provide support and encouragement in that often distressing process of moving from emotionally protective but unmanageable illusions to harsher but more manageable reality (e.g., the meditations on the loss of illusions by Wheelis, 1966).

OPERATIONS CONCEPTS: SPECIFIC

Quantification (Metrics)

There is just about nothing that brings thought down to reality like quantification! "Show me the money" and "How many (members of group X) does it take to screw in a light bulb?" are popular expressions of that idea. Metrics are highly useful in transforming the nonspecific, and therefore difficult to manage, into the specific, and therefore easier to manage.

Clinical Applications

Psychometrics is, of course, a sophisticated and highly developed field. Psychological testing can often be useful in specifying clients' underlying feelings and practical capabilities. But "back-of-the-envelope" metrification (as it is sometimes called) can also be useful.

For example, a 1-10 scale can often be helpful in assessing the appeal of an option, a ten being highly appealing, a five being middling, and a one being unappealing. Summing career opportunities 1-10, and then ranking them, something like the multiattribute utility technique (MAUT; Wright, 1984), can help sort the more appealing jobs from the less appealing jobs. The same technique can even apply to spousal candidates.

Noting *how many times* something happens in a day or a week, or noting simple *correlations* (e.g., every time A happens, B happens), can help transform vague complaints or wishes into actionable goals. As an example of useful metrics, good diet programs almost always start with counting calories, then correlating that with weight loss. And good exercise programs always keep track of distances, weights, repetitions, and times to evaluate progress. Making a habit of counting things helps a great deal to make reality more understandable and plans more manageable.

Probability

The human operations assessment of life is fundamentally *Bayesian* (Puppe, 1993; Weatherford, 1982). It means that events produced by systems (such as the things that a spouse, child, or supervisor does) are added up by the mind into track records (base rates), and those track records get translated into predictions (probabilities) as to what the system will produce (such as what the spouse, child, or supervisor will do in the future). Systems change over time, of course, and produce at different rates (e.g., children grow up, and think and act differently), so to keep up with that and stay in tune with a probable future, the mind adjusts, changing expectations to match reality.

Bayesian thinking thus could be defined as *adjusting probability estimates to keep pace with changes in the frequency of events.*

Clinical Applications

Albert Ellis specifically (Ellis & Harper, 1975), and cognitive therapists generally (Alford & Beck, 1997; Beck & Young, 1985; Burns, 1999), identified false predictions (unrealistic expectations—erroneous probability estimates) as a key source of depression. Thus, what seems like an abstruse notion from the farther reaches of academia—Bayesian thinking—can actually be seen as a core of mental health.

Forecasting and Scenarios

Probabilities always seem a little abstract, even when they are accurate and reliable. So, two related techniques from the field of operations for dealing with the future and with expectations are *forecasting* and *scenarios. Forecasting* can be defined as *a partly intuitive and partly actuarial prediction as to what will happen with people or a situation, at a key point in the future,* such as in six months or five years. Forecasts are a little less strictly mathematical than a probability statement; forecasts are as much art as science, a guestimate, a ballpark figure. *Scenarios,* then, are the *visualization of what could happen,* a phenomenological walk-through of a potential situation, a little like walking through a house that's on the market and *imagining* what it would be like to live there. (*Developing scenarios* is usually termed *scenario writing.*)

Clinical Applications

Thus, a therapist might ask a client with regard to a marriage candidate, "What would your forecast be as to the nature of this relationship in, say, five years?" And then the therapist might ask, "Shall we walk through a scenario of this relationship in five years? What do you see, what do you hear, what do you feel five years into this relationship?"

The future is a huge part of most people's lives, of course. *The future* can be interpreted as *a mental construction describing an expected reality at further points in time, including expected experiences* (phenomenology). Thus, thinking pragmatically about the future is very important, despite popular and to some extent well-taken admonitions, to live fully and perhaps even exclusively in the present moment (*be here now*, as it is sometimes put). Forecasting and scenario writing, based on adaptive, Bayesian thinking, can greatly assist in planning, arranging, and managing people's futures.

Cost-Benefit Analysis

It is a truism of decision research (Davidson, 1989) that every decision entails, from some point of view, costs as well as benefits. So, it is commonplace that the operations approach includes an analysis of the costs as well as the benefits of a course of action (Kamauff, 2010; Tomes & Hayes, 1993). This is usually summed up as a ratio of benefits to costs. A ratio of less than 1 means that the costs are bigger than the benefits. It is not that therapists and clients have to work out the ratio quantitatively. Again, it can be a back-of-the-envelope process of asking, "Altogether, do you see yourself as gaining more from this option than you are going to be paying, or is it break-even, or will you be losing?"

Clinical Applications

Clinical situations often come down to decisions—to go in one direction in life or another, to choose one option versus another. Getting into and out of relationships, careers, educational programs, or living situations flow from assessments and choices. It is easy, in the tumult of fear and desire, either to magnify the potential benefits (e.g., "It will be wonderful") and ignore the potential costs, or to magnify the potential costs (e.g., "It will be horrible") and ignore the potential benefits. A more rational cost-benefit analysis can help bring options down to earth, and can thus allow clients to make more balanced, feasible, satisfying choices.

Independence and Self-Possession

The concept of independence, as in field independence (Nunnally, 1978) or internal locus of control (Aylwin, 1985), has a distinguished history in psychology, a legacy of excellent research. *Independence* is a premise of operations—*the ability to assess situations, establish objectives, plan and initiate a course of action, and follow through to success despite distractions and conflicts*. This amounts to capable self-management.

Clinical Applications

Just as many therapeutic successes flow from caring and connection, many clinical dilemmas seem to flow from fundamental uncertainty and limitations in confidence. As clients learn to operate more effectively, they become more genuinely independent. As their independence increases, so too does their *self-possession*—their *sense of mastery over themselves and their world*. This self-possession, this sense of having an existence that is comprehensible and safe, is often what they were seeking all along.

SUMMARY

Physical operations, such as getting water and food, building a shelter, or building a relationship, are managed by the mind, by the meta-operation (Suzuki & Katayama, 1991). Familiarity with key concepts from the field of operations can help to understand and deal with challenges and problems in living.

FROM OPERATIONS TO THE OPS MODEL OF MIND

The next step in explaining operations as a basis for living and for psychotherapy is to describe the ops model of the mind in detail. This description begins with a vastly simplified yet perhaps functionally comprehensive working diagram of the mind (previously presented as figure 1.1).

Chapter 3

Human Operations Model of Mind

It might be easier to think about the mind, how it might be organized and how it might work, and easier to make adjustments to it if the mind could be *visualized*. What follows is an attempt to sketch out the structure and process of the mind from an operations perspective, beginning with the heuristic diagram presented in the first chapter (see figure 3.1).

QUICK PREVIEW OF THE COMPONENTS OF THE OPS MODEL OF THE MIND

The components of the ops model will be discussed in much detail in subsequent chapters. But it may be useful in an initial familiarization process to give a feeling for each component via the following synopses and informal definitions.

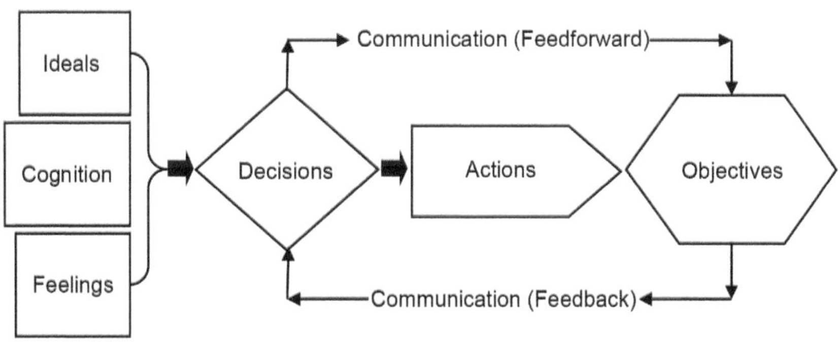

Figure 3.1 Process Model of the Human Mind.

Ideals: Component 1

The *Ideals* component includes both *aspirational* standards of conduct (*ideals*) and *practical*, everyday standards of conduct (*values*). These form the overall, upper-level *guidelines for living* adopted by an individual or a community.

Cognition: Component 2

Cognition is the *factual, logical, number-crunching* component of the mind. This is the accounting/engineering/legal/statistical department. This is the prophetic Bayesian part of the mind that provides probabilities of outcomes (Puppe, 1993).

Feelings: Component 3

The *Feelings* component includes both *sensations* and *emotions*. *Sensations* are the basic *perceptions* of the world—*color, sound, texture, fragrance,* and so on. *Emotions* then are those very distinctive *reaction tendencies associated with sensations*. Positive sensations are associated with positive emotions and approach. Negative sensations are associated with negative emotions and avoidance (or a call to the attorney).

Decisions: Component 4

The *Decisions* component is the *balanced, buoyant, enterprising, diplomatic clearing-house of the mind*, the CEO and dispatcher of the mental system. The Decisions component assesses Ideals, Cognition, and Feelings for a read on what is desired by the "self-system" (Sullivan, 1953, p. 109). Then the Decisions component generates plans that, as much as possible, are likely to satisfy the needs and desires, and then stimulates the necessary action.

Action: Component 5

Action is the concrete, observable, *behavioral* component of the mind. It consists of the tasks and steps that are hypothesized to lead to specific objectives.

Objective: Component 6

Objectives are the potential *accomplishments* that the self hypothesizes will *satisfy needs and desires*. Objectives are things people want and are willing to work for. In the ops model, the Objectives component is what everything is immediately aimed at—the target, the purpose, the *point*, the obvious "why" of doing, eventually linked to the initial want.

Communication: Component 7

Communication includes feed*back*, of course—a posteriori information as to project success, or failure, or in between. But the Communication component also includes feed*forward*—a priori *advocacy* for projects, intended to increase the probability of success (Cavanagh, 2006).

EXPLANATION OF THE FUNCTIONING OF THE OPERATIONS MODEL OF THE MIND

Overall, the *operations (ops) model of the mind is a dynamic system in which environmental input is transformed into output that assists survival and contributes to thriving and happiness*. The interactions of the seven components are primarily governed by the central evaluation and selection component (decision-making). Selections of options are based on quasi-economic assessments as to what will benefit the system (the self) and pay back the costs (time, effort, and resources). More specific aspects of the model are as follows.

Hierarchy

Effective systems are usually organized as a hierarchy (Albus, 1981; Pattee, 1973; Powers, 1973; Wilson & Gedo, 1993; Yourdon & Constantine, 1979). That way, there is an upper regulating level that helps give unity to the system. That unity increases *functionality—the ability to reach objectives*—as well as stability. That's the sports concept of "Everyone on the same page of the play book."

In the ops model, the mind is hierarchically organized. At the top of the ops system is the Ideals component. That level *dominates* (this is a technical term, not meant to endorse mean people or bad attitudes) the next level down. The Cognition component in turn dominates the Feelings component. Feeling is at the bottom of the ops system, providing basic sensory data and the emotional energy needed to fund, and carry out, specific personal projects.

This hierarchical arrangement, when well functioning, incorporates a semi-automatic process of conflict resolution. For example, powerful emotions (Feelings component) that propose a course of action that is assessed as unlikely to work, that violates probability (Cognition component, one level up), tend to be quashed, like a legal decision that violates a rule. Clever schemes (Cognition component) that propose a course of action that violates commonsense decency and fairness (Ideals component, one level up) tend to get quashed, again, like a legal decision that violates a rule. Each level, to a degree, regulates the level below it.

Left-to-Right Sequence

By convention, movement in a diagram runs left to right, and so it is in the ops model. Thinking starts on the left, with the Ideals, Cognition, and Feelings components, and runs through Decision in the center, through Action on the right, to Objectives on the far right. Information circles from Decisions over to Objectives in a feedforward (advocacy) loop. And information circles from Objectives back to Decisions in a feedback (accomplishment) loop. The left side is the Reflection side, and the right side is the Execution side. The *basic ops sequence* thus is: *Reflection–Decision–Execution*; in summary form, that would be: RDE.

This sequential, left-to-right arrangement, when functioning well, also incorporates a semi-automatic process of conflict resolution. Options need to be reflected on (on the left side) before they are transformed into active projects

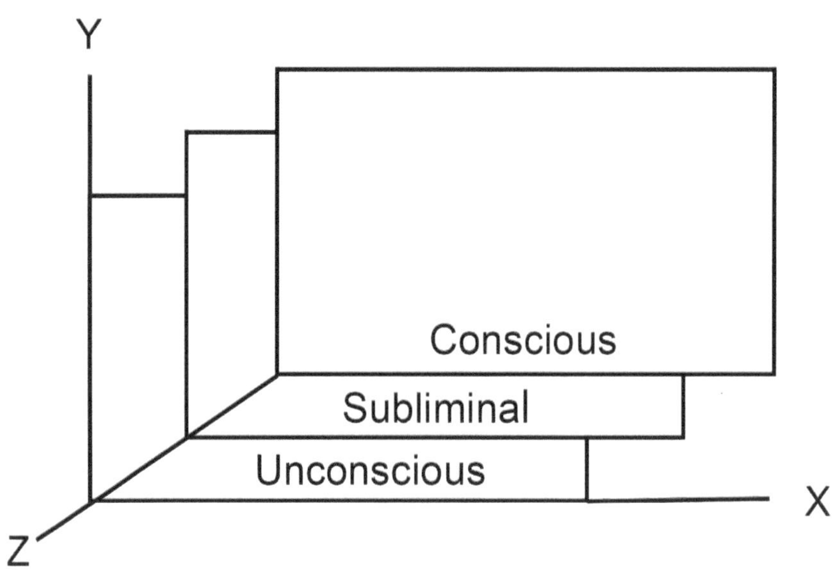

Figure 3.2 Planes of Consciousness. Conscious: Information people are directly aware of in their minds (front plane). Subliminal: Transitional region; information is somewhat accessible (middle plane). Unconscious: Deep processing and memories; obscure, difficult to access (back plane).

(on the right side). Thus, actions shouldn't be carried out impulsively; on the other hand, the system shouldn't get bogged down in paralysis of analysis, either, with ideas being reflected upon and options contemplated, irresolutely, forever.

The horizontal line running from the Cognition component through Decisions and Action to the Objectives component is, in the ops model, the critical path; this is the *productive through-line* from initial calculation to final accomplishment. In a well-functioning system, there should be left-to-right unity, resulting in consistency of perception, evaluation, and motion toward an objective—*action within a rational frame.*

Front-to-Back (Planes of Consciousness)

Figure 3.2 shows a front-to-back (z) dimension of the mind, complementing the horizontal (x) dimension, which is about focused action, and the vertical (y) dimension, which is about top-to-bottom regulation. (a) The front plane is conceptualized as *consciousness.* (b) The middle plane is conceptualized as *subliminal* mental activity: *people are typically not aware of it, but if pressed can access information in that area.* This is the transitional, tip-of-the-tongue, I-know-it-but-I-can't-say-it area. (c) The back plane is conceptualized as *unconscious* mental activity: *people are typically not aware of what goes on in that area and find it very difficult to access.* Dreams, slips of the tongue, artwork, and otherwise mysterious manifestations of the mind reveal the existence of this deep realm (e.g., Brenner, 1973), as do some neuroscience and neuroimaging studies (e.g., Dehaene, 2014; Le Doux, 1996). This *tri-planar* arrangement of *conscious-subliminal-unconscious* is an attempt to accommodate many people's phenomenological experience of their minds as well as basic determinations of psychoanalysis and recent neuroscience. The subliminal plane and the unconscious plane together in the ops system are termed *nonconsciousness.*

To enlarge on that a little bit, there is a long-standing controversy over the nature of consciousness and its relation to nonconscious realms of mind. In the ops system, consciousness versus nonconsciousness is heuristically conceived of as an information arrangement, something like the relationship between Upstairs (the owners of the house) and Downstairs (the help), portrayed in the television show *Upstairs-Downstairs* (Firkin, 1971). Commercially, an apt analogy might be the typical front-office versus back-office arrangement. In the Upstairs-Downstairs analogy, the family (Upstairs, consciousness) is busy making overall decisions, experimenting with and debating new styles and ideas, and largely interacting with the outside world. The servants (Downstairs, nonconsciousness) are preoccupied with support activity to make sure everything is running smoothly, and identifying and correcting fundamental problems. Downstairs (nonconsciousness) is to some extent detached and anonymous in relation to Upstairs (consciousness) so as to not intrude with routine or obscure matters, unless necessary. In the office analogy, the front office (consciousness) is intensely busy with phone calls, meetings, plans, and deals, while the back office (nonconsciousness) is providing nearly invisible support, archiving information, and perhaps researching puzzles in a more leisurely atmosphere. In both cases, consciousness (Upstairs, front office) is far more intense than nonconsciousness (Downstairs, back office). It may be that *consciousness* then is the *experience of that bright, buzzing intensity actively oriented to the outside world.* This may be similar to the cognitive neuroscience concept of *workspace* (Dehaene, 2014; LeDoux, 1996).

The nonconscious versus conscious distinction in the ops model seems somewhat similar to Keith Stanovich and Richard West's (2000) highly regarded distinction between *System 1* and *System 2*. The less-conscious System 1, "Operates automatically and quickly, with little or no effort" (Kahneman, 2011, p. 20), while, in a more conscious realm, "The operations of System 2 are often associated with the subjective experience of agency, choice, and concentration . . . in an orderly series of steps" (Kahneman, 2011, p. 21).

Centrality

A popular and respected concept in the study of the mind is *centering*—getting *grounded* and feeling oriented in some essential way. Mindfulness exercises often focus on centering (Suzuki, 1970). Many people do hunger for a sense of being at the heart of the present moment, in the center of their existence. In the ops model of the mind, the Decisions component, in the center, is the *pivotal component* in the system. This is where people finally live, so to speak. It corresponds to most people's sense of who they are, classically termed the *ego* (Hartmann, 1964)—that *experienced core of the self.*

Picoeconomics

Normally, people think of an economic system as being something akin to Wall Street, Hong Kong, Singapore, or The City in London, or business in general. But, as a whole, the individual and the individual mind also could be looked at as an economic system. Just as a traditional economic system is concerned about expenses and revenue, and not running out of capital, so the individual mind can be seen as being concerned with effort and reward, and not running out of energy. Benefits, over the long haul, have to outweigh costs, or the system crashes. Individuals, as well as businesses, have to be good economists to survive.

Since the individual mind is a relatively small economic system, and *pico* means *small*, the mind could be thought of as a *picoeconomic system*. Thus, thinking effectively partly means being a good picoeconomist!

INTEGRATION OF OPS SYSTEM WITH OTHER PERSONAL CHANGE SYSTEMS

The ops model incorporates numerous elements of other psychotherapy, personal change, and even management systems. It represents an attempt to provide an overall framework for some of the best-known and most respected elements of outstanding prior work.

Psychodynamics

In the ops model, the Ideals component corresponds more or less to the psychodynamic superego, the Cognition component and the critical path correspond largely to the ego, and the Feelings component corresponds to the id. Defenses, to some extent, protect as well as blind the components of the mind. In the ops model they are conceived of as a ghostly, pervasive presence that influences the functioning of all the components. In strong egos the influence of defenses is markedly reduced. Nonconscious processing is accounted for in the planes of consciousness schema. The ops model is dynamic, though not driven in the mechanistic, causal sense; rather, it is an active neurochemical information system primarily governed by decisions—selections among options according to preferences.

Cognitive Therapy

In the ops model, cognitive processes are captured in the Cognition component, the Decisions component, and to some extent in the Communication component, which informs people as to the fulfillment of their expectations.

Existentialism and Decision Theory

The central process in the ops model is decisions; making choices is the core principle of existentialism and decision theory. Existential authenticity is represented in the ops concept of *calibration*—in the largest meaning, *a functional match between inner and outer life*. (Calibration is a key concept in the ops system, and will be frequently discussed below.)

Behaviorism

Behaviorist concepts are addressed in the ops model primarily in the Action (behavior) and Communication (consequences) components. Thus, cognitive behaviorism—CBT—is also incorporated in the ops model, as the model involves both cognition and behavior.

Motivational Interviewing

The essence of Motivational Interviewing lies in contrasting the actions of the client with the statements of preference of the client. For example, many addicted clients state they would like to not be addicted, but continue to pursue their addiction—that contradiction is fed back to the clients through active listening. That contradiction is easy to visualize using the ops model—the Reflection side on the left does not match the Execution side on the right; to resolve the contradiction there needs to be a change in the middle, in the Decisions component.

Systems and Cybernetics

The ops model is a system, of course, gathering information, analyzing it, and transforming that information into decisions and action: an input-output paradigm. Cybernetically, the accomplishments, or non-accomplishments of actions are fed back into the system in the Communications component, providing environmental guidance, showing what worked and what did not work.

Management by Objectives and Management by Results

Management by objectives (MBO) and Management by results (MBR) are basic organizational concepts. MBO is about focusing on immediate achievements—the kind of thing workers and first-line supervisors can understand and focus on. MBR is the next step beyond that—if the objectives are achieved, what then? What will be the bottom-line costs and benefits to the organization? To put that in slightly different terminology—for the system as a whole, what will be the eventual *outcome*, the eventual *results*? This is similar to the excellent work on evaluating training done by Jack and Patti Phillips (1997) expressed as *return on investment* (ROI).

This distinction between initial *accomplishment of the objective* (MBO) and final *result* (MBR) is a useful one to keep in mind, and will be mentioned again in the discussions below. In the ops system, information about both the direct accomplishment of the objective, and the eventual results are assigned to the Communications component, to feedback.

It might seem a little odd to include management concepts in a book on psychotherapy. But the concept of *management* actually can be applied to systems in general, including the system that is the mind. Thus the terms stress *management*, and anger *management*. Industrial/organizational psychology, which researches management issues, actually has a lot of helpful ideas regarding perennial and very human issues such as ambition and goals, competition and conflict, frustration, and success, as well as burn-out, so it seems appropriate and useful to acknowledge management principles in the ops system.

FACTOR ANALYSIS, REGRESSION ANALYSIS, AND STRUCTURAL ANALYSIS

The ops model might appear a little oversimplified. The question might be asked, "How can seven components account for any sizable portion of human life?" But the mathematical technique of factor analysis has shown again and again that complex systems can be plausibly distilled down to a handful of core elements (Nunnally, 1978). Likewise, regression analysis and structural analysis (Cozby, 1997) have shown persuasively that many productive systems can be described with a handful of input variables. After stepping a handful of variables into a regression equation, often there is little variance remaining to be accounted for. Granted, as the British researcher of brain design, Igor Aleksander (2001, 153), pointed out, "Anything that we can think of, imagine, or plan . . . is the product of the firing of immensely complex assemblies of neurons." Nonetheless, as Douglas Riggs (1969, 2), one of the pioneers of the mathematical modeling of biological systems remarked, "It is therefore invariably necessary to reduce the complexity of the real system by making various *simplifying assumptions,*" transforming the real system into *the simplest model that could plausibly account for the system's performance*. Similarly, an award-winning cosmologist (Krauss, 1993, 25) mentioned, "On the surface, the world is a complicated place. Underneath, certain simple rules seem to operate."

CENTER POINTS, BOUNDARIES, AND LINKS: FUZZY SYSTEMS

In looking at diagrams like figure 3.1, it is natural to assume that components have crisp boundaries and that all the issues associated with a particular component are inside that component and don't stray outside, and the components have crisp and consistent links—lines of information-sharing with each other. But as the great Russian neuropsychologist Alexander Luria (1973) pointed out, the brain really isn't organized that way. Rather, there tend to be centers within the brain that carry out particular functions, but those functions are also dispersed to a degree, and everything is somewhat loosely linked. This kind of organization is sometimes referred to as *fuzzy systems* (Negoita, 1985)—there are identifiable, prototypical centers of activity (similar to the concept of centroids, or clusters in statistics), but they do not have *crisp* edges. Rather the limits, the edges around the centers, are *fuzzy*. This is not too different from the reasoning that goes into the construction of the diagnostic categories in manuals (e.g., American Psychiatric Association, 2000), in which one set of symptoms shades into another, but nonetheless there are distinct diagnostic terms and categories.

Thus, in the ops model functions are identified, and the center points of the functions are located graphically in a particular position (e.g., the Ideals component, the Cognition component), with lines connecting all. That somewhat arbitrary schematic does give the impression these are crisp entities, with crisp relationships with the other entities. But *the ops model is conceived as a fuzzy system*, therefore the ops components are conceived of as overlapping quite a bit with each other, and having various degrees of cooperative and conflicting relationships with each other in shifting configurations, sometimes sharing information and sometimes not, within a somewhat loose and variable overall command structure. This, indeed, seems to correspond generally to people's phenomenological sense of how their minds work.

RESEARCH

Many people over the years have lamented the difficulty of translating theoretical ideas into specific, testable concepts (e.g., Fawcett & Downs, 1986). If ideas are not testable, it is hard to make empirically-based improvements on them.

The ops model and the ops psychotherapy approach are intended to be tested by researchers, at least eventually. Basically, the components can be operationalized, then measured to arrive at a baseline profile of functioning. Then, change procedures can be carried out, including ops-style psychotherapy. The change procedures can be operationalized, and measured, as part of a manipulations (internal validity) check. Changes in the components can be correlated with various measures of success, such as personal satisfaction, reduction in relationship conflicts, occurrence of work promotions, or increases in income. Exploratory, even qualitative pilot studies can be followed by more rigorous studies incorporating control groups.

All of that can become fairly elaborate, of course, and expensive. However, in principle the ops model and the ops psychotherapy approach can be tested, and that testing then can provide an empirical basis for making changes in the ops formulation of personality and personality change. Appendix A and

Appendix B are a first attempt at bringing the ops ideas into the psychometric realm. Appendix A is more internal, a five-item per component self-assessment, inventorying ops strengths and limitations. Appendix B is more external, a typology, suggesting how people could be typed according to their strengths and limitations in each ops component.

CALIBRATION AND RECALIBRATION

Perhaps the key notion in the ops psychotherapy approach is *calibration—the degree to which personality components interact effectively, match personal goals, and result in surviving and thriving.* (The term is taken from the excellent work on systems by Watzlawick, Bavelas, and Jackson, 1967.) Within the ops approach, this is a holistic notion, incorporating phenomenological satisfaction in addition to more objective measures of success. *Well-calibrated* is the fundamental ops term indicating *mental health,* indicating a high degree of effectiveness, success, and satisfaction. *Mis-calibration* then is the fundamental ops term indicating *marginal or insufficient functioning,* a mismatch between components and personal goals, a low degree of effectiveness, success, and satisfaction, resulting in threats to survival. In the ops system, psychotherapy amounts to a supportive, encouraging process of *gradual recalibration.*

CHARACTER AND PERSONALITY

In the ops formulation, character and personality are related, yet distinct concepts. Both are linked to, and derived from, the seven basic components of the model.

Character is defined as *the core set of qualities that form the foundation of an effective operational personality*—such as honesty and generosity, factualness and reasonableness, compassion and affectionateness, decisiveness, initiative and follow-through, realistic ambition, enthusiasm without deception, willingness to genuinely listen to others, and an ability to strive in life within a larger frame of satisfaction. (This is the positive reverse of character disorders.) If any of these qualities are missing, those character lacunae can begin derail healthy goal-seeking (e.g., dishonest people may lose spouses, lose jobs, or go to jail).

Personality, then is defined as *the broader, more inclusive, more creative, more colorful set of qualities that go into creating a unique existence lead by a unique individual*—such as deep spirituality, an exceptional involvement in math or science, a passion for art or dance, an extensive involvement in sports, or an uncanny knack for organization, the pursuit of high accomplishment in literature or business, child-rearing or education, a flair for public speaking or music, or a fascination with the process of negotiation. These are, of course, just a fraction of the many ways that the model's seven components can combine in varying strengths to create the panoply of human personalities.

In the discussions that follow, character is treated is the more fundamental set of qualities, and personality is treated as the more inclusive and creative concept. The seven components and their arrangement and relationships within the ops model are considered, overall, to be—*the operations model of personality*. *Mind* and *thinking* are then considered as the *most* inclusive, general terms describing mental structure and function.

THE CONTEMPLATIVE LIFE VERSUS THE HEROIC LIFE

There seems to be a classic split in history and society, literature and philosophy between, on the one hand, what might be called the *contemplative life*—that dedication to savoring the beauties and fascination of life without consideration for practical benefit, and on the other hand, the *heroic life*—that dedication to engaging with the challenges of life and to triumphing over them, however difficult. The ops model of personality suggests a source for this long-standing conundrum. The left side of the model is Reflection, which may have its parallel in the contemplative perspective on life. The right side of the model is Execution, which may have its parallel in the heroic perspective on life. Both are part of thinking and living well. They are connected by decision-making. This dynamic linking of the two sides of the personality will be explored in greater detail in the chapter on action.

MODEL COMPONENTS LINKED TO INTERVENTION: A PREVIEW

The ops personality model offers a *frame of reference*, a *psychological blueprint* that the client and therapist can refer to in investigating difficulties and exploring possible personal changes in the process of moving toward greater success and comfort in life. What follows is a preview of the kinds of difficulties that can occur, and hints as to the ops approach to correcting those difficulties (explained in more detail in the next chapter):

1. *Ideals*: There may be *ideals and values* limiting the client. Perhaps some fresh ideals or adjustments in values would contribute to more success.

2. *Cognition*: There may be *cognitions* that are inaccurate, and are setting the client up for disappointment and learned helplessness. Adjustments to those beliefs may assist in improving mood.
3. *Feelings*: The client's *sensations and emotions* may be impoverished, or persistently negative, or overwhelming. The client may be cut off from basic sensory joy, or may be repressing feelings that need to be explored, appreciated, and developed.
4. *Decisions*: The client may be suffering from paralysis of analysis or over-impulsivity, or in some other way experiencing impaired *decision-making* that impedes assessment, planning, and practical accomplishment.
5. *Action*: This may be poorly-targeted, or initiation or follow-through may be insufficient to reach objectives. Corrections may be useful for greater goal achievement.
6. *Objectives*: These may be too vague or too unrealistic to be reachable. Adjustments may facilitate success.
7. *Communication*: This may be a little off-track. Either plans are not being presented with enough charismatic élan to be convincing to others, or project results are not registering sufficiently (fear of failure is too great, or even fear of success).

SUMMARY

The overall idea is that *if the client is experiencing difficulties in living, those difficulties result in some way from the functioning of the seven mental components individually or, more likely, in combination.* As those seven components are strengthened the probability of success and satisfaction in life goes up.

APPLYING THE OPS MODEL TO PSYCHOTHERAPY AND PERSONAL CHANGE

Next is a brief description of the ops psychotherapeutic process itself, primarily intended to be suggestive of how classical psychotherapeutic and other personal change techniques can be applied within an overall operations frame of reference.

Chapter 4

Operations Psychotherapy

The Process and Ultimate Goal

Probably no one would go to a doctor and say anything like, "Doc, I can't play Beethoven on the piano. Can you do something for that?" or "Doc, my golf putting is bad. Can you do something for that?" or "Doc, I can't do quadratic equations. Can you do something for that?"

And probably no health professional would assess one of these three *operational learning lacunae* as a mental health problem. And if one of these issues somehow managed to be categorized as a mental health problem, there would be no obvious way to *treat* it, or *cure* it, in the normal medical model arrangement of symptom-diagnosis-treatment.

But it may be the case that most mental health issues are actually more analogous to learning to play the piano, play golf, or solve equations than they are analogous to a previously functioning bodily system, like the lungs or the sinuses, impaired due to contracting an illness, or analogous to a machine that is jammed or broken. In other words, it may be that *most mental health issues revolve around operations that need to be mastered—goal-oriented learning issues.*

That may be hard to see because the field may be struggling under a somewhat outdated but persuasive analogical paradigm of relatively simple Newtonian mechanics versus probabilistic quantum mechanics (Hanson, 1965; Heisenberg, 1971; Krauss, 1994) and basic infection biology (Moore, 2002).

THE BEETHOVEN CHALLENGE

Music knowledge and performance are an intriguing analogy to the operations that seem to comprise much of life. Music is remarkably complex, like life. There are all those different musical notes on the staff, each note being within a key, which often change. Every note on a score corresponds to a physical movement on a keyboard or other instrument. So, it is not just *reading*; the eye and the mind and the hand have to be smoothly coordinated into an effective performance. Composers such as Beethoven often write music that is complex. And then it is not really enough to play the music technically correctly; it has to be played with the appropriate feeling and, if possible, with a little unique, bravura flair. Music teachers consistently remind students, usually with charming firmness, that there is only one way to learn music—*study and practice.*

Golf, as another comparison, is doubtless one of the most delicately complex sports ever devised. One small, seemingly inconsequential, unnoticed twist of the wrist and the ball can scream away from the fairway to splash, perhaps, into a water hazard. At Pebble Beach the ball can vanish into the ocean. Jack Nicklaus himself couldn't retrieve it. Golf pros, and even sports commentators, cheerfully remind golf aficionados that there is only one way to learn golf—*study and practice.*

As a final comparison, quadratic equations are a core part of algebra, and many people master them sufficiently to solve their assigned math problems. But everyone seems to agree that there is only one way to get to that point—*study and practice.*

The characteristics of a system's *output* can be seen as a function of the characteristics of the *system*, something like being able to perform music, play golf, or do equations as a result of learning. Related to all this, Stephen Covey (2004) wrote an outstanding guide to personal change titled *The 7 Habits of Highly Effective People*. Reading *7 Habits* it rapidly becomes clear that Covey was not so much talking about *habits*, like stopping at Starbucks every morning for coffee, as he was talking about *character*, like doing the right thing even when that is difficult.

Character, as suggested previously, can be seen as *the underlying mental organization that adaptively yet integritously supports robustly effective action*, the foundation of the more creative total personality. But the underlying character, as well as the overall personality, typically must be intentionally developed. Kernberg, Weiner, and Bardenstein (2000) detail some of the ways that character disorders can develop. Character development is first governed by the family (e.g., parents tell children to be honest, to work hard, to be kind, and so on); then the refinement and maintenance has to be taken over by the individual (e.g., dealing with later temptations and stresses). Just so, acquiring the knowledge and skill needed to play Beethoven or golf or learning to do quadratic equations can be seen as analogous to mastering life operations (e.g., how to make friends, how to successfully perform a job, or how to keep a marriage together). First, the generating system has to be developed, and then the desired output can occur.

"Playing Beethoven" can be considered analogous to almost any challenge a person might address in life; it can be considered the *paradigm operation*. That is the *Beethoven challenge* in life—*beautiful results call for extensive construction efforts, study, and practice, usually under caring guidance.* This can be a little hard to recognize because people are exposed constantly to experts (e.g., media appearances of professional entertainers, high-level executives, attorneys, and professors), who usually make the acquisition of vast stores of information and the application of graciously refined skills look charmingly easy—almost like magic.

QUICK FIXES

Humans, as a group, love magic (e.g., the best-selling alchemy novel by Paulo Coelho or the fantastically popular works about Harry Potter or the Lord of the Rings). *Magic* could be defined as *a massive positive output following minimal input of time, money, and effort.* A craving for magic might well be in the human DNA (Marean, 2015). Levers are a little magical, as Archimedes pointed out (he famously said that if he had a place to stand and a long enough lever, he could move the entire Earth). Seemingly magical, at various points in the development of civilization, are bows and arrows, pulleys, ladders, stairs, and keystones, and then steam engines, electricity, television, lasers, and mobile phones. Being able to *achieve major gains from minimal resources* (shrewd energy management, savvy picoeconomics) seems to be selected for in the competition for survival and sovereign control of resources.

It would not be surprising if mental health as a field were a little inclined toward magic. There is even a serious and fascinating book about the mental health practitioner as a magician—*The Structure of Magic* (Bandler & Grinder, 1975). The case of *Dora* (Freud, 1905/1997), alluded to in chapter 1, is somewhat similar. Dora is a complex case, involving trust and betrayal, and exploitation of youth by age evolving over a long period of time—a tragic situation, really, a cruel caricature of parental decision-making. But Freud was trying to apply a very Victorian, strictly mechanical model of science to the mind in Dora's case, rather than the subsequent fluid, more probabilistic concepts of quantum mechanics, which might be a necessary aspect of understanding semi-structured neurochemical processes, at least as an analogy (e.g., Schwartz & Begley, 2002).

In parallel research applied to the body, Louis Pasteur, a mid-nineteenth-century contemporary of Freud's, was just discovering that bacteria was the root cause of many illnesses (Moore, 2002). If the particular tiny pathogen associated with the disease could be eliminated, the disease, the whole set of symptoms, could be eliminated too. Pasteur, as well, was working within a direct, causal, simple paradigm, though in his case a biological one—simple action, huge effect.

But what if the mind is primarily a system composed of interacting levels of repertoires (subroutines) of knowledge and skills that enhance survival decisions (Albus, 1981; Tomes & Hayes, 1993; Warring, 1984)—*learned operations*? Then the kind of reductionistic, directly mechanical, externally causal, complete solutions that the discoveries of Newton and Pasteur imply may not be applicable. The appropriate approach may be an *operations-oriented learning format.*

THE OPERATIONAL VIEW OF THE MIND

If the mind is functionally composed of subroutines incorporating knowledge and skill aimed at surviving and thriving, that *operational repertoire* has somehow to be acquired (e.g., Piaget & Inhelder, 1969; Polderman, et al., 2015). For example, young children have to learn what is safe to eat and what is not safe. They have to learn when it is safe to cross the street and when it is not. They have to learn which children are OK to play with and which children could be trouble. They have to learn how to deal with their emotions—being in touch with them enough to know what they are, but not being overwhelmed by them, or, as the excellent researcher R. J. Davidson (2000) pointed out, not being haunted by feelings long after the relevance of those feelings has passed. They need to learn how to push on to completion despite fatigue. Children need to learn how to think clearly, accurately, and quickly, and act effectively, successfully. They need to learn how to operate in a multiplicity of realms large and small.

People normally are faced daily by a succession of reality demands that need to be met with reasonable felicity (Sandler & Gray, 1999). For example, they need to get up in the morning, check with the family and make sure everyone is doing OK, make breakfast, follow traffic to work without breaking too many laws, greet staff at work and renew work relationships, get the coffee, get started on the day's assignments, initiate cooperation and negotiate conflicts with other staff members, complete the day's assignments, and finally say farewell and head toward home.

When things do go wrong at work, it won't help to collapse and complain (not for too long, anyway); that is the time to regroup emotionally and tackle the situation from a new angle. When conflicts with other staff do get heated or personal, that is not really the moment to get down and dirty; rather that usually is the moment to ratchet up the diplomacy and the search for reasonable compromises (Nierenberg, 1968; Sandler & Gray, 1999).

When people finally get home after a long day, and it turns out things haven't gone well at home, people do have a little more authority and command, but after all it is a family, kids are kids, and after a searching discussion, maybe the best conclusion is lessons learned and move on. Then there is dinner, then TV or a book and popcorn, and perhaps a hopeful, bright-side review of the day, and some mental notes on making sure tomorrow will be more of a winner. That can be considered *life*, more or less!

One of the charms of the Sherlock Holmes stories is the emphasis on great learning, which leads to the clever solution of the mysteries. If the system of life operations is not learned to a reasonably competent level, then the negative emotions can start to flow in a serious way in a *phenomenology of nonsuccess*. It is often at that point that people enter psychotherapy, seeking assistance in living.

THE NATURE OF OPERATIONAL PSYCHOTHERAPY

What all of that suggests is that if clients report such issues as confusion, anxiety, depression, anger, suspiciousness, procrastination, inattention, addiction, and so on, as indicated earlier, the place to look for changes, likely, is in the functional origin—in *the personality itself* (including the underlying character). This is an indirect, second-order change approach, addressing experiential and behavioral flow by dealing with the system generating that flow (Watzlawick, Weakland, & Fisch, 1974).

There are many ways to formulate the overall personality in such a way that changes can be more easily addressed, of course (Gilliland, 1984). The ops formulation presented in this book suggests identifying seven key components and places them in an interactive model minimally sufficient to account for goal achievement and need satisfaction. The assumption is that good feelings and successful behaviors flow from a well-informed, well-organized, and well-managed personality. A well-designed personality system generates success, on the whole. By contrast, psychological difficulties are assumed to be the result of *mis-calibrations*—misalignments—in interactions among the basic components of personality or between components of the personality and the environment. The ops model allows the psychotherapist and the client to visualize the personality, assess components and interactions, and then change the components that appear to be contributing to difficulties in living.

Ops psychotherapy as presented in this book draws on a variety of established principles and techniques from both psychotherapy research and the field of operations, as well as insights from business, education, history, philosophy, and literature, and even entertainment. This helps the psychotherapist to create, in conjunction with the client, a *personal operational learning program* in which the client can make personality adjustments (recalibrations) such that life becomes more comfortable, exciting, successful, and satisfying. The goal is a stronger underlying character, a more creative and capable personality, and thereby a better life.

THUMBNAIL REVIEW OF OPS COMPONENTS

Just as a quick reminder, below are the seven basic ops components of the personality, accompanied by descriptive thumbnail sketches.

1. *Ideals*: The upper-order principles, including values, which serve as high-level reference guidelines to living successfully.
2. *Cognition*: The statistical, accountancy, probabilistic assessments that form the factual and logical basis of decision-making.
3. *Feelings*: Sensations and emotions; the perceptions and passions that form the dynamo of life.
4. *Decision-making*: The information clearinghouse of the mind; the pico-economic bottom-line selection; the best choice, all things considered.
5. *Action*: The sequence of behaviors intended to achieve an expected result; the immediate *means*.
6. *Objective*: The specific accomplishment hypothesized as likely leading to the satisfaction of a need or a wish; the immediate *end*.
7. *Communication*: The process of proactively advocating for a project, and then assessing the feedback about the success or failure of the project.

LEADING QUALITIES OF THE OPS PSYCHOTHERAPIST

The client will be coming into the office of the ops therapist looking for growth and development along operational lines. It is important that the therapist have the qualities that will be useful to the client. These specified ops qualities are common among therapists generally, but it may be useful to highlight some of the more important ones in an ops approach.

Genuine caring, as is now well known, is the foundation of almost all successful therapy. *Empathy* is particularly valuable because the ops approach works *within* the client's frame of reference, like Motivational Interviewing (Miller, Zweben, DiClemente, & Rychtarik, 1992; Miller & Rollnick, 2002). This means developing in the therapist a lively clinical *theory of mind* of any individual client. What does the client like and dislike? What does the client want to achieve and want to avoid? How would the client prefer to achieve that? What patterns of action would the client prefer to avoid? What pace of change is the client interested in? What support system does the client have in making changes? Does the client perceive conflict in the client's social system (e.g., a spouse who might be threatened by personal development), or does the client perceive collaboration? All of these things might be different in the client's mind versus the psychotherapist's mind, so it is part of the discipline of doing psychotherapy to continually return to the client's view of things, verifying understanding.

Calmness and patience are critical, because clients are often on edge to begin with and because clients often enter therapy with unrealistic expectations of ease or speed of change, and these expectations need to be discussed and negotiated in a supportive but *mildly detached way*. As the therapist *models a serene but alert and conscientious state of mind*, an analytic attitude, and gives the client a chance to explore and develop, the client has a tendency to imitate that mindset and to apply it to personal issues in a style of hopeful serenity (Hirai, 1975; Schafer, 1983; Suzuki, 1970).

Analytical ability, of course, is part of helping the client sort through issues and gain a reorientation toward true ends in life. Analytical ability here means demarcating the relevant from the irrelevant, being precise about goals and obstructions, identifying the sources of beliefs and feelings, identifying the sources of difficulties, and helping to evaluate change plans for likely effectiveness. That is, the psychotherapist needs to be a good *general problem-solver* (Frank, 1974).

And in all that, a *sense of humor* seems useful! Personal change—changing character and corresponding habits—is not unlike losing weight or getting in shape. It looks fine on paper. It is not so easy to do in real life. If personal change were as easy and fun as it usually appears in advertisements of, say, gym equipment, gambling casinos, or herbal diet pills, then everyone would become rich, thin, and marvelous overnight! But if changing attitudes, habits, and relationships is analogous to learning music, sports, or math—that is, much harder than it looks—then the process of personal change doubtless calls for a healthy dose of humor, a deep comfort with some of the absurdities of living. As the great Dutch sage Erasmus in *Praise of Folly* put it: "Many questions . . . are better ignored than investigated, seeing that part of our knowledge lies in accepting that there are some things we cannot know, and a great many more where uncertainty is more beneficial than a firm standpoint" (1971, p. 231).

OPS THERAPY: FROM OZ TO KANSAS

Indeed, there does seem to be a widespread presumption in society that a psychotherapist will somehow *apply* an almost magical technology *to the client's mind* and thereby transform long-standing attitudes and beliefs, knowledge and skills, reactions and habits, and relationships even if the client is resistant. There seems to be a myth in society that psychotherapy is something like surgery, or car repair, vaguely, a locale like Oz (Baum, 1995) where fantastic transformations can occur and realistic limits can be transcended instantly. And it is true; sometimes major changes do occur quickly, and some procedures, like medication, or hypnosis (e.g., Yapko, 1990), do seem to have quick effects. Sometimes, a listening ear and a kind, supportive word; a critical fact passed on; an insightful interpretation; or a nudge in a new direction can make all the difference. But, on the whole, the ops perspective is that recalibrating the mind usually calls for a considerable amount of focused, persistent, pragmatic effort on everyone's part over time. (This is something like *reengineering—adjusting processes to achieve different outputs*; Hammer, 1995). In other words, typically, ops psychotherapy is less like Oz, and more like Kansas (Langley, Ryerson, & Woolf, 1989).

OPS PSYCHOTHERAPY PROCESS

Recalibration of needs as well as less-essential desires emerges in the process of ops therapy. At the beginning of the therapy, it may not be clear what kind of personality development and adjustments need to occur. The learning lacunae and consequent mis-calibrations (e.g., mismatches between goals and capacities) become more obvious in the process of identifying and pursuing goals. So, ops therapy initially is fairly straightforward about goal setting and goal pursuit. The initial phase has the following steps.

1. *Active listening:* Right from the first contact, the therapist needs to establish with the client a *collaborative* process in which the client largely determines issues and directions, and the therapist regularly *checks understanding* with the client via *paraphrasing and summarizing*.
2. *Initial assessment*: Of course, there needs to be a reasonably thorough work-up of the client, so nothing important is missed—family history, education, work history, physical health, medications, suicidality, addiction history, mental status, possibly testing, and so on. Clinical psychology and psychiatric texts offer a wealth of suggestions for thorough but comfortable data gathering (e.g., Hersen, 2004; Sadock & Sadock, 2001). Linking with relevant medical personnel is always valuable. Insurance (like Caesar) may need to be dealt with, and that may involve formal diagnosis.
3. *Identification of need or desire*: *What does the client want*? That may be something concretely connected with survival, like making more money; or it may be something more experiential, like living with less anxiety,

depression, or dependence; or it may be more on the order of a dream, like finding a wonderful partner or an especially satisfying and even glamorous career. As in Motivational Interviewing, what the client really wants, deep down, becomes the leverage for resolving conflicts, as the wants are contrasted with the actions (Miller & Rollnick, 2002.)

4. *Defenses*: At every step, anxiety and conflicts can result in mental distortions, such as minimization, repression, and projection, which can misdirect the ops analysis and delay progress. That is why it is so important to generate an atmosphere of calm, supportive, objective inquiry in sessions, so that the client can relax and face issues candidly and eventually identify what is *really* wanted. (Again, patience and humor on everyone's part are invaluable!)
5. *Goal setting (end)*: As wants are identified, they can be translated into specific, operationalized objectives so that everyone will know when the goal has been achieved. This process, like most of ops therapy, is conducted in a heuristic spirit, a pragmatic *inquiry* into what the client believes will result in a satisfying life.
6. *Path determination (means)*: Here the client can state the preferred route to the goal. For example, if the initial issue is loneliness, the preferred path to companionship may be joining a professional organization where socialization is always on the agenda. If the initial issue is depression, the preferred route to a better mood might be joining a gym or looking into medications. If the initial issue is marital conflict, the preferred route might be conjoint therapy, in which marital ends and means are discussed in a supportive but candid atmosphere. The path the client chooses is often the likeliest to work, though defenses often, initially substitute misdirections.
7. *Difficulties and creative learning*: The straightforward goal-seeking process discussed above usually reveals one or more operational lacunae in one or more of the seven ops personality components. Those lacunae result in mis-calibrations (mismatches) between wants and capabilities, between preferences and probabilities—difficulties in achieving goals. The general resolution to mis-calibration is to scale back goals or change goals or, more often, to strengthen component capabilities so that goals can be reached. Strengthening capabilities usually involves designing and following a learning program—brainstorming new options, selecting the most appealing options, and then acquiring the relevant skills and knowledge step-by-step. The acquisition steps are usually specific to components (Ideals, Cognition, Feelings, etc.) and to issues (anxiety, depression, etc.), so they are detailed in the following chapters as the components and issues are individually discussed.

COACHING AND CONSULTING

Psychotherapy is a form of personal change, but so are coaching (Grodzki & Allen, 2005; Minor, 2002; Whitmore, 1992) and consulting (Block, 2000; Sears, Rudisill, & Mason-Sears, 2006; Weinberg, 1985). *Coaching* could be paradigmatically defined as *one-on-one focused teaching of key skills and attitudes needed to perform specific and general tasks. Consulting* could be paradigmatically defined as *providing analyses or other information to one or more people, allowing those people to make better decisions.* Both coaching and consulting can be seen as operationally oriented—the practitioners are trying to help people make better decisions and carry out more effective actions.

EDUCATION AND TRAINING

The landmark study on psychotherapy effectiveness by Meltzoff and Kornreich (1970) noted that, statistically, education (such as taking a class targeted to a particular issue) was relatively effective in the enterprise of personal change. And training, such as human performance improvement (e.g., Piskurich, 2002), is very much aimed at changing personal knowledge, skills, and attitudes—personal operations—in the direction of greater success. Thus, education to some extent and training to a large extent, can be seen as operationally oriented, teaching data and action patterns that are expected to achieve certain results.

Thereby, operations—means-to-ends processes—can be seen as the core of a wide range of change procedures. So, the ops approach to personality enhancement can draw on a multitude of well-developed systems and techniques in the quest for personal change.

PRAGMATIC, ADAPTIVE UTOPIANISM

Sometimes these personal change programs can seem a little ambitious, a little utopian. But, it seems like, historically (e.g., More, 1516/1961; Schaar, 1961) the downside of utopian thinking has been that it plunges into an idealized vision fully at one shot. An alternative to that is to *pilot* ideas of a better life, to approach imagined improvements *heuristically.*

Blue-skying is a motivational concept in which immediate limitations are temporarily ignored and the mind is encouraged to imagine—"What if I could have exactly what I wanted?" Once the utopian vision is sketched out, then parts of it can be *tested.*

For example, for any given individual, attending medical school might be a high ambition, a utopian dream. That ambition could be tested in a safe way by reading medical journals, by inviting doctors to lunch and talking to them about their work and lives, by frequenting a medical school and talking to the students, by volunteering at a clinic or hospital, or by talking to a counselor.

A practical and graduated approach to dreams such as this might be thought of as *heutopian thinking*—a blend of *heuristic* and *utopian*. A heutopian framing of ambitions allows clients to think about and practice living their dreams while reducing the risk of anxiety (fear of failure), failure, or depression (sense of loss).

SUMMARY

There seems to be in society, understandably, a presumption, a hope, perhaps, that psychotherapy has some kind of secret means of directly, mechanically changing the brain of the client, thereby eliminating negative psychosocial states, in a coup de maître. Medications and hypnosis, of course, are two relatively rapid routes to at least symptomatic change and often to more than that. But it may be that life is fundamentally a matter of clarifying needs and desires, establishing the plans that could reach the chosen objectives (defining means and ends), and then taking the necessary actions with sufficient follow-through to achieve success—*methodically building operational competence so that wants can be satisfied.* And if that operational competence in various areas is not acquired, then difficulties in living (symptoms) occur.

The correction to that situation may be to identify the operational lacunae—the missing knowledge and skills regarding goal-seeking—and to build the necessary operational competence step-by-step. Thus, needs can be met and desires at least partly realized, and phenomenological distress can be reduced. This is the essentially positive yet methodical process of ops therapy.

DETAILING THE OPS MODEL: INDIVIDUAL COMPONENTS

In the next section are component-by-component sets of suggestions as to how to approach personal change using the theory of mind ops.

Part II

COMPONENTS OF MIND

Chapter 5

Ideals

Personality Component 1

The tough, romantic novel *The Virginian* helped establish the Code of the West, for example, "If he gave his word, he kept it" (Wister, 1902/1979, p. x). The Virginian, far out West in Wyoming a long time ago, traveled in unregulated territory and needed a personal code of honor to guide his conduct. At the top of most systems, in a pyramidal scheme, sits a code of basic governing principles that help regulate the rest of the system and give it coherence and identity (Pattee, 1973). That top level becomes a final internal *reference level* for the system as to what is important and how to behave in order to survive (Powers, 1973). Ideals and values are the upper level of character (deep structure) and personality (surface structure).

Similarly, organizations in the maelstrom of the competitive commercial world usually have a superordinate *mission statement* and a subordinate policy and procedures (P&P) manual to which anyone seeking guidance in decision-making can refer. The mission statement and the P&P manual are supposed to keep the organization oriented in directions that are justifiable (legal as well as financial).

This pyramidal scheme of system control applies to societies as well as to individuals and organizations. As alluded to previously, most cultures have a body of myth or upper-level explanations and principles enshrined in one or more major sagas or writings that give the society a sense of identity and unity, and promulgate a vision of compassion and fairness, such as the savvy compassion of the Greek goddess of justice, Themis (Hesiod, 1973), or the grace and nobility of the Trojans, as depicted in Homer's classic heroic tale about the Trojan War (2004). Like the Virginian, Americans in World War II needed a shared code of beneficial values to make it through the war emotionally intact and created those in inspiring movies and songs, among other media (Jordan, 2000).

THE IMPACT OF IDEALS AND VALUES

These upper-level role models, truisms, and principles have an almost unfathomable impact on individuals and groups. The Middle East regularly bursts into flame over just these issues. These top-level guidelines about OK and not OK, right and wrong, permitted and denied, legal and illegal, massively influence the choices and actions underneath them. The attempt to establish fair and plausible codes of conduct can be traced at least back to Hammurabi (Bottero, 1992), to Justinian (533/1979), and to the Magna Carta (Howard, 1998). Because so much is attached to them, because so much results from these top-level principles, people with some frequency get into violent disagreements over them. A slight change in upper-level principles can have huge effects down the system, even if simply in imagination, helping some people but hurting others.

As a somewhat prosaic example, regarding urban development, Napoleon III exercised a claimed *right* (the power of eminent domain) to have Baron Haussmann cut straight through medieval Parisian neighborhoods to make those broad, modern boulevards lined with trees, lamps, and cafes (Kirkland, 2013). The demolition and reconstruction process resulted in the city looking like a war zone, very inconveniently displacing 20 percent of the population. Thus, this upper-level right of eminent domain resulted downstream, so to speak, in years of controversy and hardship. (Of course, almost everyone eventually decided they loved Haussmann's new Paris, which is still quite apparent today).

THE AMBIGUITY (AND VOLATILITY) OF GOVERNING PRINCIPLES

Meta-analysis may be relevant to the concept of upper-level governing principles. It is a statistical technique for aggregating data (Schmidt & Hunter, 1977). The idea is that research conclusions will rest on the more valid foundation of a multiplicity of studies, rather than suffer the relative unreliability and confusion of a single study that ends up being contradicted by another single study, and on and on, in a mystifying cycle of confirmation and disconfirmation. A meta-analytic basis would provide a more stable and justifiable foundation for a value system.

But, naturally, few individuals or institutions are in a position to run meta-analyses to arrive at well-grounded axiological conclusions. Therefore, it is not unusual for the determination of these upper-level ideals and values to cascade into a free-for-all, or get locked into rigid dogma, or both, because the upper-level principles have a huge impact yet remain a matter of controversy. "What is the right thing to do?" and "Therefore, what is the right decision?" are examples of the kinds of questions that seem to haunt individuals as well as communities.

All that is surprisingly similar to what the epochal Timberlawn study found about family systems (Beavers, 1977). Either dysfunctional families tend to be chaotic, with very few guiding regularities at all, or as they try to pull themselves together, they become tyrannical, a sort of peremptory family martial law, with rigid rules and no discussion. The *optimal* approach is the most difficult one—*firm but reasonable guidelines partly based on frank family discussion.*

That was the democratic approach of brilliant, classical Greece, of course, and perhaps, indirectly, of their predecessors, the peaceful, prosperous, agreeable Minoans (Fitton, 2002); thus, there are social precedents for arriving at governing principles through rational, factual discussion. But societies as well as families, and even individuals in regulating their own mental events, frequently slide into the more convenient but perhaps less healthy tendency toward various forms of tyranny.

The eminent Harvard psychologist Howard Gardner in *Truth, Beauty, and Goodness Reframed* (2011, p. 2) lamented the loss of the "classical virtues" and the current social and intellectual chaos: "Our classical virtues have been . . . pummeled by developments in our era." He concludes that people should all "assemble around a campfire" (Gardner, 2011, p. 205) and discuss these things so they can arrive at a reasonable and widely endorsed body of principles for living that provide models of generosity and responsibility.

DEFINITION OF IDEALS, VALUES, INTEGRITY, AND ROLE MODELS

Ideals in the ops system are defined as *upper-level principles that form an aspirational template for behavior*; wonderful to consider, not always easy to do. *Values* are defined as *the derived everyday guidelines for relationships, goals, and activities*—a half-level down, so to speak, from ideals. Values are the standards by which people actually live. *Integrity*, then, can be defined as *a reasonable match (calibration) between professed principles and actions.* In the ops system, that means that the ideals and values correspond fairly well to the rest of the personality, both individual personality style and the underlying character. Individuals can be seen as accountable to others for developing, or adopting, and then following through on some kind of reasonably *prosocial* code of action (or society as a whole, and thus its members, are threatened). Even the hard-boiled, fictional noir detective Philip Marlowe was an idealist at heart (Day, 2014)! *Role models* can be defined as *those individuals, groups, or institutions held in especially high esteem, considered worthy of imitation.*

RELEVANCE OF IDEALS TO THE REST OF THE OPS SYSTEM

Dysfunctional ideals and values or insufficient calibration to the self and the environment generates, as suggested above, a number of difficulties for individuals, groups, and society. The Assyrians (Leick, 2001) and Vandals (Jacobson, 2012), for example (as negative poster children), accumulated horrific, antisocial reputations without appearing to have bequeathed much in the way of compensation to the world. But the difficulties with aligning basic principles and sensible standards with behavior aren't usually as dramatic or socially destructive as that.

The usual clinical difficulty is that people with vague or contradictory ideals and values feel confused, have a hard time making decisions, don't feel very motivated to take action, and then, whatever the outcome eventually turns out to be, it's not very satisfying, because it doesn't quite match any guideline principles. Or if the selected ideals and values veer toward the antisocial, then there is a good chance the environment (other people) will take issue with those values and the resultant conduct (i.e., social punishment). That is, ideals and values overly characterized by narcissism and aggression result in difficulties for others, and those difficulties usually circle back around to the originator, as captured in the saying, "What goes around comes around." Therefore, it is likely to almost everyone's benefit, for individuals as well as groups, to effectively, successfully resolve issues with ideals and values (at least heuristically). In the ops approach that means to rationally as well as compassionately assess ideals and values, then to resolve contradictions, and then more or less to consistently follow through on the reasonably decent principles selected.

HIGH-FUNCTIONING IDEALS COMPONENT: OVERALL PORTRAIT

"Reasonably decent" of course is controversial. It helps if ideals and values can be based on *desiderata*, which could be defined as *facts and principles of broadly apparent benefit, referenced to mutual survival, wellness, and happiness*. (This is similar to the concept of utilitarianism; Mill, 1863/1987). Psychotherapeutic methods for establishing these desiderata for any given individual appear below. What is "apparently" of benefit varies from person to person and group to group. But the most important principle here is probably the first one—*the conscious attempt to look for genuine, broad benefit in life, attitudes, and action.* Not just attraction, or interest, or fascination, or glamour, or excitement, but benefit—something that is, on the whole, manifestly, reasonably healthful, upbeat, and nurturing.

For example, the notorious Weimar dancer Anita Berber started out well, studying, as a child, eurhythmics, music and dance. As an adult, she initially established a strong cabaret performance presence. But then she segued into putting on productions like "Dances of Depravity, Horror, and Ecstasy" (Brendan, 2010, para. 15), and later got sidetracked, not too surprisingly, onto drugs, criminal activities, and financial insolvency. She died at age twenty-nine, was buried in a pauper's grave, and today is almost completely forgotten.

A contrasting personality was Audrey Hepburn, who grew up in the horrors and extreme poverty of the Netherlands during World War II. She was once captured by the Nazis, who suspected her of being a child member of the Resistance, which in fact she was. She quickly managed to escape and subsequently emigrated to England. She studied dance, and then transitioned into acting. Her work ethic and screen career were legendary—a self-made woman. At the end of her life she was an ambassador for UNICEF. Her intent was to live beautifully and well, and immersed in love. She "believed in love. She believed love could heal, fix, mend, and make everything fine and good in the end" (Ferrer, 2003, p. 217). She died at age sixty-four, wealthy and famous, in a large, magnificent house in Switzerland, surrounded by family and friends.

HIGH-FUNCTIONING IDEALS COMPONENT: SPECIFIC CHARACTERISTICS

The Ideals component seems to work better when it has several characteristics:

Prosocial

The first characteristic is a fundamental linking to a positive and mutual helpful attitude, to a broad sense of reinforcement, and to mutual survival. Without this fundamental attachment to goodness in the largest sense, people seem to be characterized by a profound emotional and social emptiness; "Boundless and bare/The lone and level sands stretch far away," as put in the ironic poem about Ozymandias and lonely, vain illusions (Shelly, 1818/1998, p. 133).

Logical Derivation

The mind seems to demand a reasonably logical derivation from primary, higher-level principles to secondary, lower-level principles, so that everyday value-referenced choices have a reasonably clear connection to avowed ideals (deduction). And the mind seems to want those primary, higher-level principles to be based on a reasonably broad and sound set of facts (induction). *Illogical* principles and choices would be those that, when an attempt is made to track them back to their source, vanish in a haze of wishful thinking and free-floating premises; as the philosopher Immanuel Kant said (1781/1966), they have to have sufficient *ground.*

Clearly Articulated Content

The mind, in a way, is an ongoing negotiation process. Negotiations are assisted by clarity of position (Nierenberg, 1968). So, it helps in the internal process of issue-resolution to take a clear position. To waffle and hedge is not only uncomfortable, but makes the mind more vulnerable to paralysis of analysis, to exploitation by others, and to a reluctance to expend effort on behalf of principles that are vague. It is hard to be decisive using fundamental uncertainty as a basis. Clear articulation of basic principles helps build confidence.

Clearly Articulated Boundaries

That articulation of content means, when push comes to shove, to have a clear mind is to be willing to draw reasonable lines between the categories of OK and not OK, between acceptable and not acceptable. Learning to draw (stipulate) reasonable limits on beliefs and actions is, like learning geometry, painful, but good for the mind in the long run. The process amounts to the cultivation of *judgment*, a high and necessary skill in living well.

Consistent—Not Contradictory

Contradiction is illogic personified. Family inconsistency and contradictions have a huge and negative effect on children (Beavers, 1977). Obviously, contradictions won't go very far in court (Calleros, 2014). It is true that conflict is the bread and butter of drama: "Characters have to embody and transmit the essence of the broader conflict, to conceive and act out the pursuit of goals, experience the collisions, friction and rupture that may actually be occurring in a much wider spectrum" (Taylor, 2002, p. 84). Thereby, conflict, contradiction, and social chaos can acquire a mantle of chic and colorful sophistication. But what is exciting and entertaining on stage can be miserable phenomenologically. Borderline states, for example (Sadock & Sadock, 2001), are prone to *idealization and devaluation*—people and situations are wonderful, or horrible, or one, then the other, or both at once. Such an inconsistent and disorienting pattern of evaluation is of course highly stressful. Again, painful as it usually is clinically, in the long run, smooth mental functioning is vastly assisted if these upper-level conflicts can be worked through.

Rational Calibration

To function, ideals and values need to be calibrated to each other and to the rest of the personality as well as to the environment. Practically speaking, it will not be of much use to set as an ideal being the next General Patton in a personality of shy temperament and modest organizational talents. There may be difficulties with repression and shadow personalities if Mother Teresa is chosen as the standard in a personality that has a vigorous inclination toward dancing and late nights. On the other hand, the Wright brothers dreamed of flying, and they did accomplish that. Thus, effective calibration is a heuristic test-and-adjust process, a part of maturation, resulting in valid self-knowledge.

HIGH-FUNCTIONING IDEALS COMPONENT: OVERALL PORTRAIT

So, distilling and paraphrasing a bit, a high-functioning Ideals component in the personality has three basic qualities. First, it is well grounded. It makes sense in an obvious way. Principles can be tracked back to a valid basis. In the broadest meaning, it is *true*. Second, it is compassionate and just—healthy for all involved. People all over the world might acknowledge it as reasonable and admirable, making it *universal* and less prone to being upended or attacked. Third, it is perfectly understandable to the individual, not obscure, opaque, or occult. It is relatively easy to call to mind, to publicly explain, to personally articulate, and to act on. Thus, when acted on, it tends to be effective, to be respected by others, and to be personally satisfying. It is personally and socially *pragmatic* in the best sense.

CLINICAL INDICATORS OF DIFFICULTIES WITH IDEALS

Many people grow up in families and communities in which upper-level guidelines for living are specifically identified and clarified. They may be contained in much-loved reference books and may be hammered out in numerous discussions at dinner and in pubs, in schoolyards and residence halls, until a workable code of conduct linked to principles is established and absorbed. The best teaching, of course, is usually modeling. "Do as I say, not as I do" only goes so far.

Unfortunately, many people grow up in circumstances in which professed principles are not followed, disquieting obscurities are never clarified, arresting contradictions are not resolved, and vulnerable implausibilities are not adequately explored. Such mysterious or even dysfunctional value codes often result in a chronic sense of uncertainty, difficulty in making decisions, and difficulty in feeling satisfied by the decisions that are made. Serious confusion and disorientation at the top of the personality system has far-reaching consequences.

OPS APPROACH TO STRENGTHENING IDEALS: GENERAL

One approach to identifying higher-order principles of living is to start with a clean sheet of paper. The philosopher Friedrich Nietzsche (1886/1973) was of the opinion that with respect to values, people should consciously drop down to a neutral, pristine beginning—*a revaluation of all values*. Shunryu Suzuki (1970) reminded people that the mind can accumulate a lot of ideational and emotional baggage in life and that sometimes it is good to start over by consciously returning to a *beginner's mind*—where everything is new and refreshed. A consciously clean slate can allow people to consider ideals, values, and role models in a clear and useful way.

Many people, in the search for ideals and values, find it helpful to read materials (or even see movies or listen to music) that offer suggestions as to upper-level aspirations and guidelines in life. Some materials are widely viewed as *wisdom*, such as the blazing practicality of Confucius (n.d.) and Solomon (*New Jerusalem Bible*) and the peaceful meditations of Thomas Merton (1960). Other materials may be so ready-at-hand as to be overlooked, yet may be quite valuable, such as the autobiography of Sophia Loren (2014), a woman who rose, almost by herself, from the ashes of postwar Naples to the heights of fame, wealth, and luxury while seeming to remain inviolably human. Ideals and values have been powerfully conveyed in such materials as the movie *Casablanca* (Wallis & Curtiz, 1942), with its theme of savvy self-sacrifice, or the movie *It's a Wonderful Life* (Capra, 1946), with its theme of love and loyalty.

A commonsense approach to identifying ideals and values is to *reverse negatives into positives*. For example, would anyone want a *dishonest* son-in-law? Would anyone want a *cold, selfish* spouse? Would anyone want *indolent, violent* children? Would anyone want *messy, argumentative, potentially dangerous* neighbors? Would anyone want to rely on a job at a company that *isn't making money*? Would anyone want to live in a community run by leaders who are *irresponsible and exploitive*?

If that list of obvious negative qualities is reversed into positives, then a very quick and perhaps not unreasonable list of *virtues* pops up: *honest, warm, generous, gentle, hard-working, neat, cooperative, prosperous, responsible,* and *caring*. Clients can easily generate these lists of spontaneous preferences. The reverse of what the client doesn't want becomes the client's *protocol for a good life*. That negative-to-positive reversal amounts to an intuitive, commonsense approach to developing a heuristic set of ideals and values that likely enhance living and increase the probability of surviving and thriving for any individual.

Obviously, the bulk of most people's value system is drawn from what they grew up with, drawn from family, friends, and culture. Some of it will work and will be cherished, and passed on; some of it will not work and will be a burden, or abandoned. In psychotherapy, clients can be encouraged to access that storehouse of what might be thought of as *personal folk wisdom* and then, while dealing with the sometimes oppressive power of custom (e.g., transference), consciously *select* those elements of the personal, inherited protocol that, in the light of day, seem most beneficial and justifiable, which rationally resonate. This is a *design* approach, which is consistent generally with the ops approach of *designing a better life, a better world based on needs and desires, talents and interests*—something like design in engineering, in which traditions and innovations are blended in trial and error (Petroski, 1996).

Some clients may want the reassurance of cultural psychosocial research to refer to when considering values. Schwartz and Bardi (2001), as one example, have found a surprising amount of similarity in values around the world. Most people admire: (1) being loyal to others and forgiving their faults; (2) being responsible and concerned for others; (3) having self-direction in choosing goals and being creative; (4) showing concern for the environment and a desire to live in a beautiful world; (5) being safe and having harmonious relations with others.

In the psychotherapeutic consulting room, or in the privacy of their own homes, or with friends and family, clients can, overall, *brainstorm ideas in a cost-free, creative atmosphere*. From the brainstorming process they can *select* those ideals, values, and role models that resonate as most meaningful to them. Then they can generate their own *personal code of honor*. Identification of ideals and values is a fundamental but ongoing, exploratory, creative process (Simon, Howe, & Kirschenbaum, 1972).

OPS APPROACH TO STRENGTHENING IDEALS: SPECIFIC CALIBRATION

- *Cognition*: Good values are more reliable and impart a greater sense of security, if they are based on broad facts and careful logic.

- *Feelings*: Compassion (concern for others) and empathy (identification with others) assist in guiding upper-level principles of living toward healthy relationships with all living things, which, in the long run, enhance individual and community life.
- *Decisions*: Choices that consciously take into account both individual wants and the needs and desires of others are likelier to be acceptable to others, and are likelier to survive axiological challenges as well as outright aggression. (In a courtroom, this amounts to the rights of the individual balanced against the rights of the community.)
- *Action*: Implementing plans does frequently engender opposition. And that often does call for a certain *command forcefulness*. But that *firm assertion* is likelier to be respected and embraced if it occurs within a frame for rationality and fairness.
- *Objectives*: Purposes are likelier to be successful if they take into account, as much as possible, all stakeholders. Selfish goals are less likely to survive challenges.
- *Communication*: Effective messages are often colorful and charismatic, thereby, convincing. But persuasion that turns out to be poorly grounded in either fairness or accuracy is unlikely, in the long run, to be accepted, and that rejection then reflects poorly on the communicator. Integritous communications are, usually, ultimately the most effective.

SUMMARY

The Ideals component is the highest level of the ops personality. It includes aspirations (ideals), which may be hard to achieve in daily life, but which provide benchmarks to work toward. And it includes pragmatic guidelines (values) for everyday decision-making. The values are derived from the ideals. Both are absorbed from the family and the culture, as well as being hammered out through discussion, reading, travel and relationships, and exposure to media such as books and movies. Altogether they result in a *personal code of honor that enhances life*.

THE IDEALS-COGNITION CONNECTION

Ideals and values are relatively general principles that can be applied to a wide variety of situations. Clarity and effectiveness of the Ideals component is assisted by strong cognition.

Chapter 6

Cognition

Personality Component 2

Under the oak and olive trees, near the fountains and the Fleming Cannon, sits the columned, arched, pale Athenaeum, an elegant Mediterranean club on the grounds of one of the most unusual and distinguished schools in the world—Caltech. The purpose of the Athenaeum is to provide a lush, relaxing, elevating environment for free-flowing discussion among some of the greatest minds of the nation and the world. This represents a symbolic return to the elegant columned stoa of classical Athens, to those storied olive trees beneath the Parthenon. "There is no place like Caltech" (Goodstein, 1991, p. 17).

The great Caltech physicist Richard Feynman tried to explain the phenomenon. He said that once several universities were attempting to lure him away from Caltech. He wasn't sure what to do, since at least one of the schools was offering a huge salary increase. As he was pondering the question on the way to his office, a biologist stopped him and enthusiastically told him about an extremely important new discovery the biologist had just made about DNA. The biologist knew Feynman was a physicist, but he was so excited he had to tell someone, and he ran into Feynman, and so he told him. That convinced Feynman. He thought, essentially: "I'm staying here; I have to; I have to be in a place where people get that excited about ideas" (1985, pp. 234–35). There is a certain electrifying power to what might be called *high cognition*—relatively pure and passionate curiosity about the fundamental abstractions that describe reality. High cognition can become an ideal embedded in a culture and even in a personal protocol of living. The astonishing success of the American space program can be traced partly back to Pasadena—to Caltech and its Jet Propulsion Laboratory (JPL).

THE POWER OF COGNITION

Almost all people, not just people at the top of the cognitive range, appear to be set up to deal with reality—to carry out operations—with reasonable competence. Sometimes operational competence is easy and natural to acquire, sometimes it is not. As the psychologist David Geary (1995) pointed out, some systems of *primary* skills and knowledge, like walking, are acquired by everyone in a natural way. People can hardly be stopped from learning to walk. Other systems of skills and knowledge, like math, are the opposite. These *secondary* systems of skills and knowledge are substantially more difficult to acquire, and it is quite easy to give them a miss. Which is a shame. Abstract systems of analysis are powerful and, partly because they are difficult to acquire, lucrative. Mastery of abstract systems of analysis almost certainly separates the professionals from the hourly workers, the supervisory tier from the supervised tier, the wealthy from the poor. Therefore, mastering cognitive processes is important on the path toward successful living, toward freedom, toward self-actualization within harmonious relationships.

The astonishing classical Greeks eventually came to perceive the world as a giant dynamic configuration of elements that could be figured out because it was organized rationally, something like language (Warren, 2007). They even called the overall, beginning principle of the world *logos*, or *word*, from which the word *logic* is derived. Not surprisingly, as alluded to earlier, they poured enormous energy into developing the field of logic as a tool, an *organon*, for understanding and mastering reality (Aristotle, 2001). They were rigorously, boldly rational and speculative, seeking to penetrate to the essence of existence through debate and through the use of math and geometry, and logic.

Correspondingly, they were assiduous in the education of their young in relation to cognition—all children were expected to attend school, pay strict attention to the teacher, practice their letters and sums, memorize poetry, learn music, engage in physical training, and, in general, strive for personal and social, heroic excellence: *arête* (Spivey & Squire, 2004). Classical Greece traditionally stands as one of the pinnacles of human achievement in the arts, philosophy, and science, as well as wealth and military prowess. This was probably not by accident. Their success may have flowed primarily from the precise, intentional development and application of outstanding cognition. And the demise of classical Greece may have resulted from a deterioration of that same disciplined cognition (Polybius, 1979).

The relevance of the brilliance of classical Greece to individuals, organizations, and societies today is that, fundamentally, the classical Greeks constructed and then drew on a system of mental development that remains valid and accessible in today's world. This book is partly an attempt to recapture that intentional, passionate, action-oriented style of thinking—that heroic style of operating informed by highly capable cognition.

DEFINITION OF COGNITION COMPONENT

Obviously, the average person, on a quest toward exemplary cognitive functioning, doesn't need to reach the heights of Caltech and JPL, or classical Athens. But polishing cognition has a tendency to move people smartly in that direction.

Cognition, within the ops system, could be defined as *the factual, logical, probabilistic, objective recognition and assessment of reality, both external and internal*. Good cognition has good reality contact and can be relied on to provide relatively valid descriptions of objects, people, and events.

In the ops formulation, cognition, like the other components, runs from front to back, along the z-dimension. So, some cognition occurs in the subliminal plane, and some occurs farther back, in the unconscious plane. Therefore, all cognition is not immediately available to consciousness.

RELEVANCE OF COGNITION TO THE REST OF THE OPS SYSTEM

As is well known, brains are locked inside, and protected by, skulls. Minds are the *experience* of brain activity. Perceptual systems (visual, acoustic, proprioceptive, etc.) *inform* the brain as to what is going on in the world. That information, once organized, in the ops formulation primarily becomes cognition. Good decision-making ultimately rests on good cognition. And good cognition requires good reality contact, which is not necessarily that easy to accomplish.

And likely there are many reasons for that. First, for example, as suggested previously, reality seems to be semi-chaotic, with lots of changes going on all the time; therefore, reality is actually not that easy to identify and structure. Second, as many people are all too aware, if some stimulus array is colorful, or loud, or new, that can end up biasing cognition in favor of the exciting but trivial and transient. (Science has not yet determined that there is such a thing as a *tabloid gene*, but such a discovery would not be surprising.) Third, emotions have a powerful effect on minds. If something is frightening, even if not quite dangerous, or highly desirable, even if not quite useful, that can register in the mind as having outsized importance, and thus bias cognition.

Within the ops system, the Cognition component largely functions as a (presumably) trusted *advisor* to the Decisions component, an interior consultant. This, analogously, is something like the role played on Star Trek's Starship Enterprise by the hyperrational Mr. Spock in relation to Captain Kirk (CBS, 2015). Captain Kirk would routinely ask Mr. Spock for one of his factual, logical opinions as to the nature of a situation or its likely outcome, and Mr. Spock would supply the data. With that information, the captain would then make an operational decision.

Cognition, in the ops model, collects and catalogs incoming information, analyzes it, and then draws conclusions as to the probability of danger or of safety in a given situation. Cognition functions, to provide another analogy, as a precise, internal *operational handbook as to what works when.*

HIGH-FUNCTIONING COGNITION COMPONENT: OVERALL PORTRAIT

People who function well in the area of cognition usually have an intuition that their thinking is strong. They can see it in their ability to solve problems and reach goals, and they can hear it in a positive feedback from parents, teachers, friends, and later, employers.

In the ops model, the positive paradigm of people with good cognition is that they are sharp and quick, confident in their assessments, and usually accurate. They enjoy using their minds and find working on challenges, on difficult problems, intriguing and satisfying. They are rigorous in their thinking, preferring precision over vagueness, specificity over ambiguity, and numbers over generalities. They are largely objective, evidence based, and categorical, even as they recognize that there is a certain irreducible mystery to life as well as almost artistic nuances to human relationships and striving. They are curious and often enjoy pursuing ideas for their own sake. Nonetheless, their ultimate preference tends to be to turn those initial curiosities into well-understood models (handy summations) that can be applied to practical problems. They work hard to build organized bodies of relevant knowledge (cases) and repertoires of skills, which allow them to rapidly understand situations, resolve obstructions, and reach goals. A certain game-like attitude helps their cognition stay light-footed, alert, and objective as they carry out the operations of living.

HIGH-FUNCTIONING COGNITION COMPONENT: SPECIFIC CHARACTERISTICS

Good cognition has a long and storied history, but within the ops system, the following characteristics of success-oriented cognition are considered to be of special importance. The characteristics are accompanied by various suggestions as to how the characteristics might be strengthened with the personality.

Operational Intelligence

IQ is the great and intimidating icon of capable cognition. But, indeed, the standard IQ test (Wechsler, 2008) is not only interesting in its own right, but indirectly instructive in how to think. The Wechsler runs through a series of questions covering a critical set of intellectual areas, such as memory, vocabulary, and math. Despite the many, often valid criticisms it has accumulated over the years, this test corresponds well to the assessment results looked for by schools, industry, and government. It provides hard data on education and problem-solving capacity. If the test questions are reversed and summarized, it becomes a general guideline for psychosocial students and professionals in productive cognition.

One of the long-standing criticisms of the traditional conceptualization of intelligence is that it resides in something of an ivory tower—sheer thought detached from any practical, adaptive considerations. Some efforts at bringing the idea of intelligence down to pragmatic ground have been the concepts of *multiple intelligences* (Gardner, 1993), *behavioral intelligence, practical intelligence*, and *conation.* The renowned psychologist Robert J. Sternberg (2012) has addressed the functional adaptation issue in the concept of *successful intelligence.* The concept of an operational mind may assist in this enterprise of adequately defining and assessing intelligence, in that the concept of operations incorporates observation, evaluation, planning, execution, and adaptation to results. This system of abilities might be termed *operational intelligence—the degree to which means and ends can be expeditiously selected and linked in support of individual and group survival.*

Numerous sources available to a wider public discuss and offer exercises to improve cognitive capacities. A fairly typical example, *Blueprints for Memory* (Hersey, 1990) gives many suggestions for remembering such key information as names, faces, and dates, for example—mnemonics, alphabetizing, rhyming, visualizing, and repeating. Solid and reliable materials for learning logic are, of course, readily available, such as the concise *Being Logical* (McInerny, 2004) and the fun *Logic for Dummies* (Zegarelli, 2007). Lyle Bourne and Alice Healy (2014) have produced a mini-training handbook, published by the American Psychological Association, *Train Your Mind for Peak Performance*, which addresses the speed of acquisition, retention, and generalization, among other issues. It is replete with exercises to polish cognitive skills.

Steve Martin (2001, p. 383) thoughtfully suggested this recall exercise. "Sit in your garden and try to remember your dog's name. Here's how: simply watch the dog's ears while calling out pet names at random." (He appears to have been kidding, however.)

Curiosity

Pure interest (high cognition) in how things work and why things happen can have exemplary results. A classic story (Moore, 2002) of observant curiosity having huge payoffs is the well-known discovery by a laboratory genius named Fleming. He noticed that some kind of mold (*Penicillium notatum*) was messing up the petri dishes and accidentally killing what was supposed to be growing in the petri dishes (*Streptococcus*). It appeared as if the mold mess needed to be cleaned up. "Wait!," was the genius thought. "*Something . . . is . . . killing . . . streptococcus*." And thus penicillin was born.

Nurturing sheer curiosity, in oneself or others, such as children and students, can be highly useful. Great discoveries don't always have obvious payoffs in the initial stages. Focused self-talk can often help in modeling and developing mental capacities; for example, phrases such as "This is interesting . . . I wonder how this works" or "I wonder what was the cause of this" can help direct thought toward a creative form of pure curiosity.

Explicit Situation Identification

Similarly, in the rush of the day's activities and the flood of information, there can be a certain genius to stopping and registering what is *actually* going on. Lawyers are trained to think with lightning speed (Calleros, 2014) yet to simultaneously pause and think, "OK, what do we *really* have here? What happened, and who is this, *exactly*? What do they want, *specifically*? What did they say, *literally*?" Explicit (yet efficient) identification of object, event, person, and statement is a useful habit, which can be developed through conscious intent. This is similar to the journalists' mantra of who-what-when-where-why-how.

Definition of Terms

In the same vein, definition is surprisingly powerful. The landmark Scottish philosopher David Hume (1748) as well as the Nobel-prize-winning physicist Richard Feynman (1985) said that if some statement (e.g., a book or a theoretical proposal) cannot be understood after a reasonable amount of effort, the fault usually can be traced to inadequate definition.

There are many ways to approach definition, of course. But one way, drawing on fuzzy systems theory (Negoita, 1985), is to conceptualize a useful definition as having three parts, culminating in a tangible referent. First, there is a *boundary* around the issue or entity, creating a kind of category, even though the boundary might be stipulated (a little arbitrary). The boundary (the edges of the category) demarcate what is in and what is out, what is included and what is excluded, what is relevant and what is irrelevant. Second, there is a *core prototype*, identifying whatever the thing is in essence. The classic example of a prototype is a robin, which exemplifies the bird category. Even though the category includes penguins, which swim and don't fly, penguins remain within the boundary of the category, while robins anchor the category. Third, there is a *label* capturing the entity, like "birds." These three criteria amount to the key idea of *tangible referent*, or set of referents, that exemplify the term "physical circumstances" as Feynman put it (1985, p. 245). This is similar to the industrial/organizational concept of *criterion-referencing* and related to the scientific measurement concept of *operationalization*. Careful definition results in a *there* there—something actually can be found when listeners and readers arrive at the point to which language has directed them.

Classification

Taxonomies, such as educational taxonomies, can be intimidatingly complex, massive, and unwieldy for many purposes. Nonetheless, putting things into groups, which are then put into a level system, is not only amazingly useful but basic to thought. Not surprisingly, classifications are ubiquitous (Upton, 1941). A corporate organizational chart is an example. Military ranks amount to a classification. James Albus's (1981) conceptualization of the underlying organization of robots is hierarchical, comprising levels and groups of outputs—a classification.

It is not that hard to get in the habit of automatically classifying things. The trick is to ask, "What is this thing a type of?," and then, a step further, "What if I put these types together—what would be the supertype, the type of the type?" In other words, it is amazing how much of reality can be grouped meaningfully, and once people know what *type* of thing they are dealing with, they can handle situations more confidently and fluidly.

For example, meeting someone socially might seem like a particular situation, and a job interview might seem like a very different situation, but they both belong to a type—*an initial informational meeting*. Once the similarity within the type is recognized, operational competence can be more easily transferred.

Deduction

Deduction, mentioned earlier, is an amazingly slick and solid way of arriving at conclusions. Deduction revolves around identifying characteristics within groups. Then, any member of that group likely will have that characteristic. A famous deduction is that Socrates eventually will die because he is a member of the group *human*, and all members of that group share the characteristic *mortality*.

The way deduction can play out clinically is shown in the book *Women Who Love Too Much*, by Robin Norwood (1985). Norwood says that reasoning about boyfriends too often goes something like this: this man acts like a dream date; only a dream date would act like a dream date; therefore, this man must be a dream date. But Norwood makes a logical counterproposal: this man acts like a perfect dream date; normal men act normal, but men who are dishonest act like perfect dream dates; therefore, this man is not a perfect dream date; he is a dishonest person.

Classical logic (Aristotelean) can seem a little clumsy because it seems to depend on the concepts *all* or *none*. But it actually works almost as well if something like *most* or *almost all* or *90 percent* is substituted for *all*, or if something like *10 percent* or *very few* is substituted for *none*—that is, if tight certainties are transformed into looser *probabilities* (Kotz & Stroup, 1983). With percentages, deduction becomes a far more flexible tool of thought. For example, if, say, 85 percent of depressed people are troubled by false beliefs, and if Client A is depressed, then Client A *probably* is troubled by false beliefs.

Induction

Induction, also mentioned earlier, is the art of drawing conclusions based on a survey of evidence. Induction actually forms the basis of deduction, since the premise of a deduction (the starting point in reasoning) is actually a conclusion based on evidence.

This logical process is a core of CBT (cognitive behavioral therapy)—basing conclusions on a reasonable evidentiary basis. Depressed and anxious people can be seen as having marked difficulties with induction (e.g., "All people must always like me") as well as deduction ("If not, then it will be a catastrophe").

Mastering induction and deduction forms a powerful cognitive combination. Together, induction and deduction carry individuals and groups a long way down the path toward thinking well and feeling good.

INDICATORS OF CLINICAL COGNITIVE DIFFICULTY

Often, clients complain of conclusions that do not quite seem to have a basis but are upsetting, or facts that do not quite seem to fall into place, or confusion about who people are and what they probably will do. This amounts to an uncomfortable degree of social disorientation based on false data. In such cases, it may be that the cognition component could benefit from strengthening.

OPS APPROACH TO STRENGTHENING COGNITION: GENERAL

As usual, probably the key element in a change program is recognizing that there *is* a limitation and making the commitment to improving things. Focus usually follows that. The focus can be on, say, curiosity and the creativity that often accompanies curiosity. Clients can encourage themselves with targeted positive self-talk, such as, "Hey, this is really interesting—I think I am going to follow up on this." Or the focus may be on getting more explicit about people and relationships. Time can be set aside, say, on a Sunday morning, with a cup of coffee and pen and paper, to get clear about the nature and intent of other key people and the actual quality of relationships with those people. If definitions seem a little vague and thus misleading, clients can follow the one-minute rule and take 60 seconds from time to time to really clarify what they mean by pivotal concepts such as *happiness, love*, or *success*. Life can seem overwhelming, but it can be comfortably organized if it is classified into levels of importance. Things grouped at the lower levels (like closet cleaning) are trumped by things at the upper levels (like relationships). Priorities often flow from classifications. If upsetting beliefs seem like they might be a little illogical, clients can engage in a reverse *if-then* logical exercise: "I know what the *then* is, I know what I believe, but what is the *if* . . . *why* do I believe that? What is the *data* that supports the *conclusion*?" The alarm of catastrophizing often fades away after a careful, reversed if-then analysis. Clients may feel discouraged that their cognition is not all that they would like it to be. A reminder of the Japanese *kaizen* principle can reassure them that *life is a learning curve* and that, if they focus on it, they can get better and better, year by year at living rationally and happily.

OPS APPROACH TO STRENGTHENING COGNITION: SPECIFIC CALIBRATION

- *Ideals*: The upper-order role models and values of the mind are in a way the final guidelines of the personality, an internal *court of last resort*, as they say in Hong Kong. It is exactly the nature of ideals that they are a little removed from, and in a way *safe* from the *Sturm und Drang* of ordinary living. But for that same reason they can be a little out of touch with reality. Sound cognition—a solid grasp of reality—provides rational weighting to the Ideals component.
- *Feelings*: The Scarlett O'Haras and Rhett Butlers of the mind, however charismatic, can and do go over the top regularly. Again, as with ideals, sound cognition provides a fact check for those colorful and electrifying notions that sometimes end up, well—gone with the wind.
- *Decisions*: Good choices depend on good facts and reasoning. The Cognition component is conceptualized as having something of an automatic braking relationship with the Decisions component—facts and logic to a degree *constrain* decision-making. But mostly the Cognition component is

conceptualized as a sort of consultant to the Decisions component. CBT is, in a sense, a process of constantly checking with cognition as decisions are made about mood—whether to be happy or unhappy, elated or discouraged.

- *Action*: Cognition provides the facts and probabilities that form the basis of strategy and the planning of action. Cognition specifies what actions, in what order, are likely to work. Strong cognition helps control impulsivity and disorganization.
- *Objectives*: Choosing goals is partly a matter of fun and feeling. (Or what is a life for?) Nonetheless, it is not very much fun to go tilting at windmills over and over. Good cognition helps insure that objectives will be reached and will be satisfying.
- *Communication*: Good communication calls for a certain creative flair, a certain daring insouciance, even in a court of law. But effective communication ultimately rests on empirical data and sound reasoning. That a statement is found to be *accurate* (well grounded in facts and logic) is its own best argument.

SUMMARY

The Cognitive component can be thought of as that subsystem of the mind that collects data about reality and organizes it logically. It provides a detailed template for what is going on and for what will probably work in any given situation. It is the actuarial advisor within the mind.

THE COGNITION-FEELINGS CONNECTION

Feelings can be thought of as a *coefficient of cognition*. Sensations provide much of the raw data for cognition. An active, sensory lifestyle gives color and texture to thinking. Strong emotions and high excitement can, under the best circumstances, contribute to focused analysis and reasoning. Ambition and desire contribute to intellectual goal setting and learning. Therefore, a plan to improve cognition needs to include a program aimed at the rational cultivation of feelings.

Chapter 7

Feelings

Personality Component 3

"The Force, Luke . . ."; the movie *Star Wars* (Lucas, 1975) probably wasn't meant to be a textbook on operations psychology, but it works remarkably well as that. Perhaps because George Lucas, the creator, drew on hero myths for his saga—classic, timeless stories from around the world that draw deeply from the human psyche (Campbell, 2004).

There is a dynamic, dialectical interplay between the youthful, and to some extent unbridled, enthusiasm of Princess Leia, Han Solo, and Luke Skywalker, on the one hand, and the serene, secure, almost spiritual wisdom of Obi-Wan Kenobi and the presence of the near-mystical Force, on the other hand. Yoda, the enigmatic Jedi trainer, then functions as something of a cool, cognitive intermediary, blending Luke's raw, youthful emotional energy with the mature, idealistic elevation of Obi-Wan's wisdom, transforming Luke into a savvy but responsible, tough but noble Jedi Knight (Lucas & Kershner, 1980).

The three-level conceptualization of Obi-Wan, Yoda, and Luke corresponds to the three levels of the left side of the ops personality model—ideals, cognition, and feelings. An interpretation of the initial *Star Wars* saga is that youthful enthusiasm (Luke: Feelings) is not only necessary for life but perhaps is the beautiful, touching essence of life. However, it needs to be guided overall by a higher wisdom (Obi-Wan: Ideals). Still, that is not enough to cope with evil (Darth Vader: negative Cognition). But not just any training—the training depicted in *Star Wars* has to shape the mind and body into a focused whole that responds quickly and confidently to danger and opportunity within a larger framework of self-respect, caring, and accountability, creating an intrinsic nobility. This is similar to the *ops ethos* presented in this book.

DEFINITION OF FEELINGS COMPONENT

In the ops system, *feelings* is a composite term that includes *sensations*, such as color, odor, texture, and taste, as well as *emotions*, such as love and desire, anger and fear, happiness and sadness. The ops interpretation of the relationship between sensations and emotions is that sensations provide fundamental information as to pleasure, gain, and safety versus pain, loss, and danger—a positive versus negative message in relation to survival. Sensations are the basic signals as to the nature and status of reality; those are the *perceptions* (external and internal) that form a large part of the foundation of cognition. So, *feelings* could be paradigmatically defined as *sensations and emotions that link the mind to reality, power action, and provide the almost artistic, subtle shadings of existence.*

Emotions then are primarily the energy that needs to be associated with the sensation for the signal system to work. If the sensation is positive, emotional energy is almost immediately directed toward the stimulus; there is some kind of *liking*, resulting in active and perhaps vigorous engagement—a recognition that "*This is good.*" If the sensation is negative, energy is almost immediately directed away from the stimulus; there is some kind of *disliking*, resulting in active and perhaps vigorous disengagement—a recognition that "*This is bad.*" Phenomenologically, emotions add a fantastically nuanced, profoundly *moving* aura to the stimulus—thrilling, terrifying, deliriously attractive, disgusting, enraging, and so on—through the almost infinite palette of emotional variation. But emotions almost always distill down to positive/good/like versus negative/bad/dislike. And they are usually linked to action (approach or avoidance), often in a powerful, immediate way so as to take advantage of perhaps fleeting positive opportunities or to escape perhaps unexpected and immediate dangers.

RELEVANCE OF FEELINGS TO THE REST OF THE OPS SYSTEM

The mind is blind without sensations, of course; sensations are the necessary, constant input as to shifts in interior and exterior reality—hot or cold, hard or soft, sharp or dull, acidic or sweet, moving or stable, and so on, information essential to successful adaptation. Thus, as suggested previously, sensations provide the raw material, the raw data, for the rest of the personality to use in survival processes.

And that seems innocent enough. It doesn't seem as if feeling *good* would be a *bad thing*. But sensations, along with emotions, have had a checkered career in relation to the rest of the system, particularly in relation to ideals (e.g., Blom, 2011; Freud, 1930/1961; Nietzsche, 1886/1973). The fundamental reason, perhaps, is that each component of the mind is *semi-independent.* The esteemed, pioneering Russian neuropsychologist Luria (1973) concluded that the brain is organized that way, with capable subsystems carrying out proprietary projects with relative vigor and dispatch, within a certain amount of freedom of operation. This is a little like a distributed management system—the offices out in the hustings, so to speak, to some extent do things their own way, to match local needs. The Timberlawn study of family systems (Beavers, 1977) found that, similarly, happy families are characterized by a free-flowing, semi-organized, cheerful energy aimed at prized objectives.

The difficulty with feelings is exactly that they are one of the mind's most direct connections to reality. If something dangerous happens, the mind needs to know about it and be able to respond *immediately and powerfully.* To be motivating, a good thing, a benefit, may need to have an almost overwhelmingly galvanizing appeal if danger is to be escaped. For example, if someone is trapped briefly under water, air needs to feel persuasively good; or if someone gets far too hot, cooling shade could be a lifesaver, so coolness needs to be viscerally appealing; or if someone is actually dying of hunger, even bad food needs to taste fabulous, enormously attractive, and compelling, thus lifesaving. So, survival means that feelings, good and bad, must have *substantial immediate authority within the system,* something like an ambulance screaming down a street or the honking of a food truck that arrives unexpectedly in a crowded and chaotic refugee camp.

However, just as everyone in *Star Wars* kept a weather eye on Han Solo because he radiated a particularly jocular, feisty defiance of constraints in general (Lucas, 1975), the personality has to exercise a certain caution with regard to feelings if they are not to run wild. Both desire and fear (e.g., mania and addictions, panic and paranoia) can be dangerously transporting if they are not managed astutely. Their power and immediacy are both an advantage and a difficulty. The attempt to manage the sometimes emergency impact of feelings often results in hypervigilance and over-control. The Feelings component, like a talented and passionate but somewhat unstable member of a family, calls for *careful*, but not *excessive*, management; *discipline*, but *not harsh* discipline; *control*, but not *over-control.*

HIGH-FUNCTIONING FEELINGS COMPONENT: OVERALL PORTRAIT

The *Diagnostic and Statistical Manual of Mental Disorders* (*DSM*) of the American Psychiatric Association (e.g., *DSM-II,* 2nd ed., 1968; or *DSM-IV-TR,* 4th ed., text rev., 2000) shows a long-standing distinction between *normal disturbances* of feeling, such as grief, and actual *mental disorders*, such as clinical depression. In other words, it is not realistic to expect perfect moods.

Life has its ups and downs, its successful days and its disappointing days, and *emotions largely parallel the nature of those events*. It is normal and motivating (reinforcing) to feel good when things go well, and normal and motivating (punishing) to feel bad when things don't go so well. Feelings, as sovereign and electrifying as they are often experienced, are a functional *part* of a larger, survival-referenced system of monitoring and consequencing. Feelings are a *component* of the personality. Gaining that perspective is a necessary aspect of the maturation process. Thus, feelings are wonderful, on the whole, and important, but can't be allowed to run the system, any more than a cruise director (the hospitality and entertainment manager) would be allowed to navigate a cruise ship. What is beautiful and feels right is a *touchstone* within the system, basic and critical, but still an *informational and energizing component* of a larger survival strategy.

Overall, within the ops perspective, a high-functioning Feelings component would result in people largely enjoying sensations and emotions, and experiencing their feelings as *making sense*, as the phrase goes. On the whole, if things go well, people feel good. When things don't work out as planned, disappointment is experienced, but not as something crushing, rather as a painful learning process. Sensations are enjoyed, generally. And emotions are enjoyed, generally. Unpleasant sensations are recognized and dealt with accordingly—drinking water when thirsty, coming into shelter and out of the rain or a too-hot sun, relaxing when fatigued, and so on. Unpleasant emotions are faced and an attempt made to deal with the underlying situations—a failure and disappointment calling for greater effort in the future, perhaps; a rift in a personal relationship calling for an apology and reconciliation, perhaps; and so on. *Feelings and associated actions are relatively straightforward;* vagueness and unresolved contradictions, mysterious origins and opaque intentions, avoided realities and neglected responsibilities do not characterize their lives or minds, and therefore, they rarely experience the negative feelings associated with disorienting, out-of-control, self-defeating habits of mind.

HIGH-FUNCTIONING FEELING COMPONENT: SPECIFIC CHARACTERISTICS

Because feelings can be so incendiary and volatile, so blazing and colorful, historically it has been difficult to conceptualize them as a necessary and functional part of the personality and to manage them skillfully. It has not been hard to see them in a negative light and to find them sometimes simply pernicious and intractable. So, it may be useful, from an ops perspective, to showcase specific qualities of feelings that seem to contribute to a high-functioning personality.

Clarity

It is good to know *what is* being experienced, in the sense that if someone asked, the sensation or emotion could be identified (as during an interview with a doctor or within a relationship).

Vividness

The power of feelings is important—more powerful feelings contribute to stronger motivation. The knack of *experiencing positive feelings strongly and negative feelings less strongly* can be learned by individuals and taught by parents and teachers. That acquired knack results in feelings that are strong, but largely positive—*joie de vivre.*

Loyalty

In a strange way, the *heart* of people's lives are feelings—perhaps because feelings are so directly connected to survival. Without some kind of powerful sensory or emotional connection to the world, people often feel lost, or empty. Therefore, it seems important for people to establish deep and genuine connections to other people, places, missions, and dreams if they are not to feel like the drifting and dangerous existential stranger described by the novelist Camus (1942).

Adaptiveness

Having said that, it is also important for people to exercise judgment about feelings and to let things go that are no longer meaningful, or are beyond hope (R. J. Davidson, 2000). Sometimes sadness and stress are best left behind (which is different than repressing them).

CLINICAL INDICATORS OF DIFFICULTIES WITH FEELINGS

Most clients are fairly direct about the fact that they are having trouble emotionally. That is typically the main reason they come to therapy. Naturally, often there is a long and tangled trail from the emotions initially presented to the eventual picture of true, underlying mood and emotions linked to often partially disguised preferences and plans, what might be called *private operations*, similar to the family systems concept of *family secrets*. Clarifying all that is part of the process of psychotherapy.

OPS APPROACH TO STRENGTHENING FEELINGS: GENERAL

The overall ops approach to difficulties with feelings has two phases. The first, preliminary phase is primarily for the therapist and perhaps for therapist trainers—to consider adopting, or enhancing, a conceptualization of life and people as essentially good; difficulties as understandable; and improvements as possible even where not certain—giving therapy a *positive aura* (e.g., Fredrickson, 2009), thus boosting hope while leveraging the placebo effect. The second, clinical phase is to listen carefully to the client, of course, to connect with deep feelings and sometimes unspoken thoughts, and then to clarify desired ends and preferred means: *underneath* or *behind* the obvious feelings—what is it that the client *really wants* (e.g., as discussed by Horney, 1950; Miller & Rollnick, 2002; Schafer, 1983)?

Even if clients prefer not to share their private operations with multiple parties—their semi-hidden world of goals and dreams, wishes and fears—it is usually helpful for the clients *themselves* to know how their mind is actually working. As clients clarify genuine preferences and intentions, the therapist can draw on various change agent roles—consultant, coach, guide, cheerleader, and so on—to help clients overcome fears and discomforts and to reorient activities and relationships to form a better match with the true self.

OPS APPROACH TO STRENGTHENING FEELINGS: SPECIFIC CALIBRATION

- *Ideals*: An emotional mismatch with the top level of the personality is a prescription for misery. It is important to be realistic in selecting ideals, role models, and values, reaching neither too high nor too low. If set too high, clients may always feel a pervasive sense of wrongness, of inadequacy. But if set too low, clients may end up living a life of questionable thoughts and actions, marginal vocational and social position, and low self-esteem. In either case, depression is a threat.
- *Cognition*: Cognition has a tendency to regulate emotion to a degree. Facts and logic have a tendency to form a phenomenological counterweight to both questionable enthusiasms and nearly baseless fears. "It's not *that*

great" or "It's not *that* scary" might be typical counterbalancing comments cognition expresses in the presence of unrealistic feelings.

- *Decision*: Because the Decisions component is the functional clearinghouse and overall adjudicator of the personality in the ops model of mind, it is important for the Feelings component to give due respect, as it were, to that overall *magisterial* role, even when sensations are intense or emotions are running high, unless the situation is a true emergency. That cultivating of the Feelings component by the Decisions component, nurturing as well as guiding it, is a delicate and difficult process, calling for much judgment, but may be a core of the concept of *maturation*. It would be unhealthy for feelings to be phenomenologically bullied or completely ignored, but it would be equally unhealthy for feelings to run roughshod over sound decision-making.
- *Action*: In anxiety, the trick is to start slowly and comfortably, easing into a project a step at a time, desensitizing until confidence is reached. In depression, the trick is make the first step highly appealing, requiring minimal initial effort in seeking to pair positive feelings with action, so that effort becomes intrinsically reinforcing, experienced as—"This is working! I'm doing well!" Action is difficult for people of sensitive and emotional temperament, so it needs to be organized and initiated in a comfortable, manageable way.
- *Objectives*: In dealing with anxiety and depression, setting objectives needs to be restricted, visualizable, and well within reach, to *maximize the probability of experiencing success*. Overall, there needs to be a reasonably good match between ability and ambitions, or emotions may get crunched. Most people are equipped to operate effectively someplace in the available range of social strata. Emotional difficulties, such as depression or anxiety, can emerge if ambitions are set too high or too low. Few people have the set of high abilities called for in top executives, senators, surgeons, or first-rank artists, for example. If clients assume they have that degree of ability, and get lost in impossible dreams, they may become quite stressed. On the other hand, if clients are highly talented, and do little with their abilities, they may eventually feel a sense of loss, a painful longing for lost opportunities, as well as losing out on the rewards that normally follow the blending of talent and effort.
- *Communication*: When feelings are dicey, as they are with many clients, it is especially important to be cautious in feedforward—minimal claims and modest plans. Negative feedback from self and others should be considered, but the emphasis initially should be on looking for and emphasizing positive feedback, to build confidence in carrying out operations. When feelings are under relatively good control (no excessive depression or anxiety), then it does make sense to pour positive emotion into dreams and projects,

to pump up enthusiasm, and thereby enhance the probability of success in an upbeat program of feedforward. In dealing with sincere feedback (not a disguised attack), it is usually best to consciously dampen emotions, to let the facts and opinions emerge so they can be assessed rationally.

SUMMARY

Does anything in the human mind have the wildly variable reputation of feelings? Feelings seem to be the deep and cherished heart of human life and also the notorious villain in countless human tragedies. The explanation may be that sensations keep people in close contact with reality (which can be highly beneficial or highly dangerous), and those sensations are intimately linked to powerful emotions, neurochemical processes that have to virtually explode in times of critical opportunity or danger. So, on the whole, emotions are on our side. However, that same semi-chaotic nature requires, during maturation, a certain liberal and understanding acceptance, coupled with firm and shrewd self-management if feelings are to be enjoyed, but kept in perspective, within a larger framework of plans and relationships.

THE FEELINGS-DECISIONS CONNECTION

Sophia Loren, in her brilliantly sensible autobiography (2014), was at pains to explain a particular paradox of living, the paradox of love and money. It is good, even essential, to enjoy people and events genuinely and deeply for their own sake (something like the I-Thou concept of Martin Buber, 1970). Nonetheless, she emphasized (as in Buber's It-World) that to live a good life it is also necessary to think clearly and factually, in a practical, self-respecting, self-caring way about the development of marketable abilities, about building concrete, step-by-step success to arrive at ultimate financial security. In the ops system, that partly means linking feelings with good decisions.

Chapter 8

Decisions

Personality Component 4

If people are perfectly subject to causal forces, then other people ought to be able to *make* them do exactly what they want when they want. But they can't. Here is perhaps the most famous, most incendiary myth characterizing the power of humanity to make choices—*Prometheus* (represented in New York City by the gold figure holding a torch at the Rockefeller Center).

Prometheus was the famous Titan who took fire from its celestial sphere and brought it as a warming, enlightening gift to humanity. The gods (being all too human, so to speak) were upset. So, they bound Prometheus to a rock as punishment. Then along came Hermes, messenger of the gods. Hermes said, in effect, "Look, no one is out to get you, really. But you have misbehaved. I have it on good authority that if you will simply apologize, and admit that you were wrong to help humanity, you will be freed." Now, if there was ever a situation when someone could be *forced* to do something, it would appear to be this. Eternal punishment, or easy escape. But Prometheus was *above* being forced—he was *already* free, in a higher sense. He said to Hermes, "Understand this: I would not change my painful plight, on any terms, for your servile humility" (Aeschylus, 1961, p. 49).

There is a persistent phantom suspicion in psychology and philosophy that there must be a little person inside the head making the decisions. It is not true, of course, though the idea continues to exert, somehow, a certain plausible appeal. And, indeed, it does seem like most systems have some kind of a centralized decision-making *process*. As indicated earlier, James Albus's (1981) scheme of robotic organization has decision-making radiating from the top, like a classification. If robots are making benefit-referenced, if-then selections, decision-making does not appear to have to be conscious, which is part of the confusion. Indeed, probably most human decisions aren't conscious. That is accounted for in the ops model by planes of consciousness

(Figure. 3.2). Decisions in the first, front plane are conscious. But the other two planes behind, the supporting planes, encompass what is subliminal and then, more deeply, what is unconscious in decision-making.

For example, many people have had the experience of reading while waiting for a friend to arrive to go out and get pizza, or something like that. The visitor is supposed to arrive at, say, 6:00. Though the reader is deeply immersed in, say, a mystery novel, lo and behold, unexpectedly in the conscious plane, at exactly 6:00 the reader perks up, looks around, and checks the door. That can be seen as subliminal decision-making—tracking time internally, then a decision transitioning from the supporting, nonconscious realm of the mind into active consciousness to check for the visitor at exactly the right moment. At a deeper plane, a spouse may, for example, arrange a marital fight *in order to* justify storming out of the house, roaring out of the driveway, absently driving around, and then coincidentally ending up at a favorite sports bar the night of a playoff, drinking beer and playing pool—all nonconsciously guided and, being nonconscious, easier to defend.

A basic idea in the ops system is that there is some kind of centralized, evaluative clearinghouse of information overall that assesses options in reference to needs, desires, and other criteria and selects the option that looks the most favorable based on all the relevant data available. It is something like looking at cars, and considering price, appearance, reliability, mileage, and so on—multiple criteria, something like in multivariate structural analysis—and then making a selection. Some of it is objective (the buyer might ask, "What was the mileage again?"), some of it is subjective (the buyer might comment, "It's a beauty"); it all goes into a semi-chaotic mix, and out comes an answer (the buyer says, "Yes, we'll take it" or "No, we'll wait").

There is an old conundrum about decisions and responsibility in philosophy, in law, and in society generally that goes something like this: Is humankind *determined*, or *free*? It is a colorful and arresting question. But like many classic conundrums (like Zeno's beguiling paradoxes), it is a little bit of a play on words, a little bit of a conceptual simplification that results in a dramatic, alarming contrast. The essential line of questions goes something as follows: Are people not a part of the physical world? And if they are part of the physical world, are they not subject to causality? So, if their decisions are causally determined, how is it that people can make choices freely? Then, if they are not making choices freely, how can they be held accountable for their decisions, which have been determined by external events?

The conceptual sleight of hand in this conundrum is the suggestion that the mind is a simple causal system, something like a hammer hitting a nail. But that, of course, is a gross simplification and distortion of the way the mind works. The mind is immensely, almost unbelievably, and perhaps uniquely complex. It is far from being an ordinary hammer-and-nail, force-on-object, immediate-impact-to-result, causal arrangement. Rather, incoming information is assessed, reorganized, perhaps even radically transformed, and then retained in various forms in the memory. Decision-making is then a remarkably subtle, complex, shifting process of selecting among alternatives, based on recall, interpretation, moods, current circumstances, alternations in allies and competitors, and hypothesized future costs and benefits, among many other variables. Though based on information, decision-making can be seen as a fundamentally *strategic process* typically aimed at achieving some kind of advantage—qualitatively different from a simple, mechanical, causal process. "Strategy . . . is about decisions that matter" (Freedman, 2013, p. xiv); *strategizing* has a long history in the human saga. The often-observed fact that, as with the Prometheus myth, no one can actually *force* (as in physical force) someone to do something, supports the idea that people ultimately make choices, and thus are accountable for them.

That, at least, is the ops interpretation of the situation. That interpretation is freeing for the individual, in that people don't have to feel they are merely victims of fate or destiny; rather, they have great latitude in choosing the personality they want to have and the life they want to lead. And that interpretation is freeing for society (*other* people), since accountability is partly what encourages people to respect and accommodate to others. Thus, society, too, does not have to feel it is the victim of unseen, intractable forces acting through an individual. (This choice-therefore-accountability formulation, of course, stands in contrast to those electrifying and spooky reports about people who were forced to kill because of too much sugar, too much money, too much neglect, too much opportunity, or even too much trauma.)

DEFINITION OF DECISIONS COMPONENT

Decision-making could be defined as *the process of selecting a preferred option from an array of considered options in relation to goals, context, and overall welfare.* This appears to be an essentially rational process, in which expected costs are balanced against expected benefits. Irrationality would be a *violation of utility*—essentially, a violation of common sense. There has been a fair amount of consternation in the past few decades over the possibility that human decisions, including economic decisions, are irrational frequently and persistently enough that it calls into question human rationality (e.g., Davidson, 1989; Kahneman, 2011). But it may be that this interpretation is somewhat illusory, in that studies can be set up that showcase the flaws in people's reasoning, and then reasoning as a whole can be considered flawed. However, if decisions were too irrational, it seems unlikely people would have survived. On the whole, people seem to make choices in friendships, education, career, marriage, and self-management that largely work to their benefit—are rational (Baron, 1985). Human decision-making may partake of Simon's (1976) concept of *satisficing*—good enough is good enough, not perfect, but it gets the job done.

RELEVANCE OF DECISIONS TO THE OTHER OPS COMPONENTS OF THE MIND

In the ops model, the Decisions component is the overall governing component—storm-center in the human mind. Schematically, information is fed into the Decisions component from the components on the left—Ideals, Cognition, and Feelings. In addition, the Decisions component can access Ideals, Cognition, and Feelings in the process of assessing and selecting options. This is a back-and-forth, *internally interactive* process. Then, based on needs and desires, plans are developed involving specific means and ends—the Action component and the Objectives component. There is also a back-and-forth, interactive process between the Decisions component and the Action and Objectives component as strategies and tactics (on the right side of the model) are designed, tested, and adjusted to increase success in satisfying wants identified (from the left side of the model).

HIGH-FUNCTIONING DECISIONS COMPONENT: OVERALL PORTRAIT

From an ops perspective, people with a smoothly functioning Decisions component generally have a good idea of what they want and how to get it. They are unafraid of facing the facts that underlie a good decision, are willing to draw obvious conclusions, and are unafraid of facing their own feelings with regard to an issue. They are comfortable contemplating the right thing to do, as well as judiciously considering the practical thing to do, and make an effort to blend the two. They have the self-possession to review their preferences and options before making a choice and are perfectly willing to consider other people's relevant opinions. They have little trouble selecting a particular objective and planning how to reach it. They understand all decisions are heuristic, and, while staying flexible and making every effort to be successful, they are willing to accept failure, if it occurs, with minimal loss of self-esteem or motivation.

A neuropsychological examination, like the *Luria-Nebraska Neuropsychological Battery* (Golden, 1980), could be thought of as a *sequence of decisions* as to what the correct response would be to the series of questions and instructions. People with excellent neuropsychological functioning tend to run through that sequence of decisions with *quick, precise, fluid confidence.* Posttest interviews suggest that such people believe that making effective, efficient decisions is important, and they continuously try to improve their skill.

HIGH-FUNCTIONING DECISIONS COMPONENT: SPECIFIC CHARACTERISTICS

The characteristics of a high-functioning Decisions component are similar to the characteristics of a good therapist. The relationship between the Decisions component and the functioning of the rest of the personality is analogous to the relationship between the therapist and the client's issues and efforts to resolve them.

A Sense of Being Centered

As suggested previously, a core concept of Eastern thinking is being a calm center. This serene centering is assisted by meditation, of course (e.g., Hanh, 2010; Hirai, 1975; Suzuki, 1970). But a part of what helps that serene centering process is simply consciously cultivating a sense of inner certainty, a profound sense of belonging and OK-ness.

Analytic Distance

In a related concept, the esteemed psychologist Roy Schafer (1983) proposed that to function well the mind needs to be *engaged* with reality, while not being *overwhelmed* by reality. A way to do that is to cultivate an *analytic attitude*—in touch with events and feelings, while riding slightly above them. This concept is something like a balanced surfer moving across roiling waves or like a captain on the bridge of a ship navigating across the water. The idea is to blend passion and concern with a certain detached, objective appraisal, attaining a functional balance of closeness and distance, so as to not be swamped by the various feelings and thoughts that often rush through the personality.

Game-Like Approach

The economic theorists John von Neumann and Oskar Morgenstern (1953) and the psychiatrist Eric Berne (1961) discussed games in human life. The term *game* strikes many people as trivializing life. But, interestingly, as alluded to previously, a game can be understood as a kind of operation. Therefore, if life is an operation, life is something of a game—basically, it has rules guiding moves leading to objectives, which results in a win or loss relative to a competitor. A common connotation of the term *game* is that it is not real life, it usually has no real consequences, and *therefore* it can be entered into with a certain gusto—a shrewd aliveness to opportunities, a racy willingness to take risks, and an admirable philosophicalness in the face of both failure (e.g., "We'll get them next year") and success (e.g., "Well, that's it—let's go get a pizza"). Thus, that *game-like approach*, within reason, can be helpful in cheerfully and skillfully managing the operations that largely make up life. That approach can help free anxious and depressed clients from some of the tension and grief associated with the struggle to cope with the goal-seeking process.

Brainstorming and Creativity

Operations are by nature sequential and logical—*logistical*. Therefore, it is crucial to build into the decision-making process an open-minded willingness to consider all options, at least periodically, in a freewheeling process of idea generation and evaluation. What might seem at first like *wild ideas* can then be cogently evaluated and, if promising, added to ongoing logistics. The brilliant Steven Jobs of Apple and iPhone fame had a regular Monday meeting at which everything and anything could be reviewed from a fresh perspective (Isaacson, 2013).

Good at Absorbing Information

The Decisions component is conceived of as being in the thick of all the mental and physical activity, and to be effective needs to be *alert and retentive*. That *mindfulness* coupled with *extensive but flexible recall* facilitates efficient, sound decision-making. Alert retentiveness is a characteristic that can be consciously cultivated, the opposite of a laissez-faire, careless attitude.

Good at Identifying and Assessing Options

Life can go by in a blur, so it is useful to learn to recognize relevant options *as* options, to be able to say, for example, "Hey, *here* is an interesting opportunity" or "Uh-oh, *that* looks a little too risky." Once identified, the MAUT, as briefly mentioned in chapter 2, can be useful in ranking options in relation to desirability (Wright, 1984). The basic idea is to evaluate all the characteristics of an option, say, on a 1-10 scale, add up the scores, and average them. A used car might look like this: Price = 8, Condition = 4, and Appearance = 3. Added together, that makes 15. If 15 is divided by 3, that gives that used car a rating of 5. When done with several cars, they can be ranked—and the highest-ranking car is selected. MAUT is surprisingly applicable to numerous types of decisions—from schools and careers to houses and relationships.

Good at Prioritizing

As information from various sources is filtered, preferred ideas begin to take shape, intuitions start to turn into objectives, and possible paths start to turn into projects. That art of transforming intuitions into focused activity and then into achievements is one that can be refined through conscious practice, especially with assistance. As options are ranked, priorities can be developed. A good categorical scheme for priorities is: A = critical, B = supportive of critical activities (but could be dispensed with), and C = completely discretionary (i.e., fun or interesting, but irrelevant to critical goals and activities) (Lakein, 1989).

Good at Monitoring

Once projects are launched, they need to be monitored. Again, a good way to do that is to apply, in good operational style, some kind of a metric. If it is a weight loss program, for example, progress can be easily monitored by getting on the scale. But even the course of relationships can be assessed week to week on a ten-point scale, to see how things are going. Steven Jobs regularly monitored activity at Apple Corporation partly just by wandering around, looking things over, picking things up, and chatting with people (Isaacson, 2013).

Flexible Persistence

Once projects get going, enthusiasm can flag, unpleasant and unexpected difficulties can arise, and other people can withdraw their support, among other problems. It is primarily up to the Decisions component to keep going until success is achieved. On the other hand, ideas that seemed great when they started, like a particular relationship or training program, can start to look like a bad bet after a while. In such cases, sometimes the most sensible decision is to just drop it. It is a judgment call, and though other people can be consulted (e.g., friends, therapists and coaches), experience is probably the best teacher.

Savvy about Defenses

Realistically, psychological defenses (e.g., Vaillant, 1977; Wheelis, 1966) are almost certainly not completely avoidable. It is irritating to find, in hindsight, that defenses (such as denial or projection) have distorted insight and injured relationships and prospects. On the other hand, defenses do, to some extent, buffer the ego from the ongoing stresses of reality. Perhaps the most sophisticated approach is for clients to strive to consistently level with themselves and others in a nondefensive manner, but to accept, in a sage if rueful way, that sometimes defenses do interfere with living well and to not stress excessively over those little lies they tell themselves.

DECISIONS COMPONENT AS A SHOWRUNNER

As indicated at the beginning of this chapter, there is, of course, no *little person* in the brain that makes decisions. But, then again, it does seem like the Decisions component is remarkably similar to the position of a *showrunner* in a television show (Murphy, 2014). This analogy may be useful, if it is not too misleading.

A *showrunner* could be defined as *the serene, cheerful still point within all the chaos; the creative vision behind the production; the sure hand that guides creative possibility to artistic success on time, on budget; the one who makes everything work.* The showrunner is the boots-on-the-ground individual responsible for, and in charge of, the total production—in and out of the studio, on and off the set, on and off the phone, in and out of meetings (Murphy, 2014). The showrunner is an amazing blend of left- and right-brain functioning—logical, yet artistic. As an analogy, applied to the mind, that means the Decisions component keeps track of critical events in the self and in the environment, and orchestrates the seven components such that goals are achieved, without getting unduly unbalanced or distracted by the constant, kaleidoscoping shifts in reality or by the omnipresence of risk and competition.

CLINICAL INDICATORS OF DECISIONS COMPONENT DIFFICULTY

There are two main kinds of difficulties with decision-making—*impulsivity* and *paralysis of analysis*, derived from the two basic processes of the mind—*initiation* and *inhibition*. So, the two main kinds of clinical complaints are that the client jumps, so to speak, at options without thinking them through, sometimes getting lucky but often suffering decision regret. Or the client is perpetually torn between options in an approach-avoidance war of attrition and has a hard time making a choice; that stasis then often results in losing out on opportunities and slowing down or postponing success.

OPS APPROACH TO STRENGTHENING DECISIONS: GENERAL

There are two general aspects to improving decision-making in the ops system. The first is for the client to *recognize* the pervasive advantage of good decision-making. Without decisions, operations come to a halt. With decisions, operations can flow more smoothly. The second aspect is *practice*. Overcoming impulsivity or stasis or both is partly a matter of making a conscious effort to repeatedly *make measured, considered decisions in a timely fashion*, thereby reducing the associated anxiety to a manageable level.

OPS APPROACH TO STRENGTHENING DECISIONS: SPECIFIC CALIBRATION

- *Ideals*: The Decisions component is in the middle of the ops critical path—on the firing line, so to speak. Nonetheless, it is good for individuals to learn to regularly check in with values so that the decisions made are fundamentally *right* and, personally, deeply satisfying, rather than just *effective*, so that any later regrets and embarrassments are minimized, not to mention angry responses from others, leading to tarnished reputations. (That is, *results* need to be balanced with *goodness* if people are to fully prosper.)
- *Cognition*: This is the primary link in the decision-making process. There is no point in pursuing doomed strategies. The Cognition component provides the solid data that supports the strategies and tactics that lead to the acquisition of the supplies necessary for living. It takes a while to hone reality checking and decision-making into a smooth, flexible, synergistic relationship.
- *Feelings*: Decisions and feelings have a somewhat paradoxical relationship. Decisions often need to override feelings (as in going jogging on a cold, early morning or pushing away chocolate). Yet decisions need feelings for the colorful enthusiasm and energy of project creation, engagement, and completion. If there is little enthusiasm, there probably will be little action. Therefore, it is helpful to cultivate an enthusiastic *lieutenant* (support) role for the Feelings component in relation to the more commanding, *magisterial* Decisions component. Similarly, it is helpful to cultivate a firm yet appreciative, even *artistic* attitude on the part of the Decisions component with respect to the Feelings component, something like a schedule- and budget-minded showrunner dealing with talented actors.
- *Action*: As the book *Execution* pointed out (Bossidy & Charan, 2002), there is little benefit to plans that can't be executed; on the other hand, if plans are good, then things do need to be happening on the basis of that plan.

Boot camp exercises at 7 a.m. on the beach may look good on paper, but not so appealing in execution. Developing a mutually adjusting relationship between planning and action strengthens both. As plans are developed, behavioral techniques can be applied to chosen operations—graduated stages, prompting, shaping, reinforcing, and so on.

- *Objectives*: The Ideals, Cognition, and Feelings components provide thoughts as to needs and desires—what is *wanted*. Then the Decisions component formulates an objective that likely will fulfill the need. This process of translating often vague intuitions and impulses into concrete objectives, this *funneling process*, actually can be somewhat difficult and even painful, as various kinds of dreams are brought down to reality. But practice makes it much easier to translate wants into feasible, motivating projects.
- *Communication:* With regard to feedforward, decision effectiveness is usually enhanced by developing the habit of thinking positively about choices (cognition) and backing them with enthusiasm (feelings). There is something of an art to promoting ideas persuasively while not exaggerating them, or misleading or pressuring others about them. That art can be developed with conscious, consistent effort, as well as savvy guidance (e.g., Hauge & Vogler, 2003). Then, too, with regard to feedback, every decision is a hypothesis, so that good decision-making entails learning to expect and deal honestly but flexibly with periodic, inevitable disappointments.

SUMMARY

Decision-making is the center of the ops model of mind. Selecting means and ends options is the essence of an operation. Intuition of needs and desires is the input to the Decisions component, and corresponding plans and actions are the output, leading, when successful, to the satisfying of wants. A plausible analogy to the Decisions component is the concept of a showrunner—a blend of artistry and logistics that keeps the personality moving forward through the kaleidoscoping changes in reality toward success.

THE DECISIONS-ACTION CONNECTION

Action is what keeps plans grounded—seeing if the plans can actually be carried out. Action tests ideas. Action is where hopes start being realized.

Chapter 9

Action

Personality Component 5

Raymond Chandler's mystery *Farewell, My Lovely* (1940/1992) crackled with sun-bleached, austere action descriptions of the sleuth Philip Marlowe roving the mean streets of Los Angeles: he should have gone home, but he had things on his mind, so he kept driving east on sunset, then turned north at La Brea, then rolled over to Highland, through Cahuenga Pass and down Ventura Boulevard, through Studio City, Sherman Oaks, Encino. And that was just driving around. When he finally got into a bar, "I jumped around behind the counter. A sawed-off-shotgun lay under a towel on a shelf under the bar. Beside it was a cigar box. In the cigar box was a 0.38 automatic. I took both of them" (Chandler, 1940/1992, p. 15).

The hard-boiled American detective story from the mid-twentieth century is powerfully, crisply focused on action. These riveting mythologies of tough decency were popular then and remain so today.

Chandler's fictional hero Marlowe, searching for justice, is similar to the mythological Greek hero Jason, who famously searched for the Golden Fleece in northern Anatolia (Apollonius, 1959). At dawn on the day of departure to Anatolia, Jason, the brave, strong captain, strode with his war gear out of his house, through the town and the crowds down to his sailors and to his ship, the *Argo*, at the harbor. Then, not unlike modern partying heroes, Jason held a barbecue, with dancing, feasting, and wine, surrounded by admiring men and women. Finally, Jason and his *Argo* crew briskly left the shore behind, crossed the gulf, and headed out to sea in a stiff breeze, zigzagging along the coasts on the way to the Golden Fleece, and to adventure.

THE REFLECTION-EXECUTION (APPARENT) CONFLICT

There seems to be something riveting about *crisp execution*, about succinct, targeted *action* that resonates with people as fundamental to life and success. At the same time, it is clear that profound *reflection and planning* are important in living intelligently. Not surprisingly, a huge bifurcation in philosophy going back thousands of years, from India to China to the West, can be exactly seen along that reflection-execution split.

Indian Tradition

The fascinating *Rig Veda* (O'Flaherty, 1981) of ancient India radiates a certain bouncing confidence, as in, "Whatever I boldly set out to do, I do" (p. 169) and "Let me obtain great riches of glory" (p. 180). But by contrast, the later *Upanishads* (1965) strike a contrary note, suggesting that true wisdom and freedom from discomfort come from an "austere harmony," "by contemplation" (p. 101); "In peace beyond desires he watches the drama of the universe" (p. 100).

The famed *Bhagavad Gita* (Sinha, 1986) of India may have started out bold and focused: "In this path . . . there is one-pointed determination; but the thoughts of the irresolute are many-branched and endless" (p. 193), and "Discipline yourself in . . . achieving excellence in action" (p. 195). But by contrast, later versions of the *Gita,* perhaps out of political caution, promulgated a contrary passive approach to life that suppressed "creative and rational thinking" (p. 109).

Chinese Tradition

Confucius (n.d.), a proponent of practical behavior, said, "By his earnest activity, his achievements were great" (p. 300) and "The superior man is modest in his speech, but exceeds in his action" (p.210) and "See what a man does" (p. 16). But by contrast, Lao Tzu (1963, p. 59) persuasively, if somewhat opaquely, argued the opposite pole, promoting a wait-and-see attitude, and then adapting to whatever happens: "Do that which consists in taking no action"!

Western Tradition

And, finally, as a perhaps outstanding example, the philosopher Nietzsche (1886/1973) railed at length at how a Dionysian, dancing exuberance that is in some way the essence of living well ends up being squelched by an overly composed, retiring, and over-accommodating frame of mind.

OPERATIONAL INTEGRATION

So, all that contrast between serene contemplation and decisive boldness has been a little confusing. An explanation, perhaps, is that not only are life and the mind *combinatorial*, they are also *contingent*—what is a good idea at an initial phase may not be a good idea at the subsequent phase; what will work under a first set of circumstances may not work under a second. That *phase-based thinking* is accounted for in the ops model by the initial Reflection side to the left of decision-making and the subsequent Execution side to the right of decision-making. Both, as suggested previously, are needed for good thinking and functioning. The eminent quantum physicist Werner Heisenberg (1958, 206) suggested it is reasonable to, "hope that in the final state of unification many different cultural traditions may live together and may combine different human endeavors into a new kind of balance between thought and deed, between activity and meditation." In the initial phase of dealing with an issue, it is good to stay calm, get a little distance, reflect, and consider options. Some decision research (e.g., Payne, Braunstein, & Carroll, 1978) suggested, as did Lao Tzu (1963), that surprisingly often the best decision is no decision at all—better to mull it over while seeing how the situation plays out. Likewise, as the Palo Alto systems group pointed out (Watzlawick, Weakland, & Fisch, 1974), so-called "solutions" all too frequently make things worse.

But, finally, the *Argo* has to leave the town if it is to get to the Golden Fleece, so to speak. At the point where something has to happen, something has to happen—there has to be a shift from the Reflection side to the Execution side. As Napoleon admiringly said about Caesar, "What speed! What dispatch! What boldness!" (Bonaparte, 1992, p. 28). *Action* is that golden point at which *ideas* begin to be realized.

Perhaps the strongest and most effective personalities are those that consciously cultivate both sides of the mind. Emblematic of that proposal in the Eastern tradition is the prototypical samurai, who was expected to be focused, brave, and energetic, but also sensitive, thoughtful, and artistic, a lover of painting and poetry (e.g., Musashi, 1974; Yamamoto, 2008; Yuzan, 1941/1988). Emblematic of that proposal in the Western tradition is Alexander the Great, who was a focused and ambitious leader, but also, following the guidance of his tutor Aristotle, became a lover of books, literature, and philosophy (Plutarch, 1973).

DEFINITION OF ACTION COMPONENT

Action in the ops system could be defined as *a sequence of behaviors intended to achieve a particular objective in the service of a larger goal.* The Action component could be considered an amalgam of *behavior management, project management, and time and materials management* within an overall framework of operations. Action, finally, is where things get done, where dreams are realized.

RELEVANCE OF ACTION TO THE REST OF THE OPS SYSTEM

Actions need to fit within the larger personality so that what is accomplished leads to genuine satisfaction, avoiding the kind of potential regret captured in the famous reminder, "For what shall it profit a man, if he shall gain the whole world, and lose his own soul?" (Mark 8:36, King James Version). The overall goal is not just action aimed at survival, but living aimed at a larger, generative sense of *meaning* (Erikson, 1979).

HIGH-FUNCTIONING ACTION COMPONENT: OVERALL PORTRAIT

High-ops people are confident in taking action. They feel comfortable on the firing line and have faith in their ability to read situations and other people and in their ability to make the right move when the time comes. They are in sync with the *rhythm of operations—reflection, selection, execution*. They do not feel compelled to take immediate action, though they can take quick action when necessary. For example, General George Patton's writing (1947) is a fairly good model for this well-informed and well-reasoned yet crisp, bold, decisive style of thinking. High-ops people enjoy being able to review available options until the best one can be selected. Once an objective is selected, they are able to quickly put corresponding tasks into a sequence and to begin at the appropriate step. Once projects are begun, they have the persistence to expeditiously push those projects through to successful completion, despite fatigue, boredom, or unexpected difficulties. They carry out their tasks with a certain invigorating flair, understanding that this is their life. Then, success achieved, serious celebrations are the order of the day.

HIGH-FUNCTIONING ACTION COMPONENT: SPECIFIC CHARACTERISTICS

Behavior management, project management, and time and materials management are all about effective action—how to get *operationally organized*. Behavior management techniques (e.g., Thoresen & Mahoney, 1974) help people *get* going on and *keep* going on things they want to do or know they should do, making sure the cueing and consequencing are appropriately designed. Project management (e.g., Dinsmore, 1990; Kendrick, 2004) is an amazingly clever and massive methodology—the kind of thing NASA does—for sequencing and monitoring goal-oriented tasks; it is especially useful for complex, budgeted operations with multiple deadlines (e.g., replanting the entire yard, organizing a child's birthday party, or planning a shopping trip with multiple stops). Time and materials management techniques (e.g., Jones, 2010; Schlenger & Roesch, 1989) help prevent chaos, help put everything in order so key items are available at a moment's notice.

As Pepper White (1991) said about his student years at MIT, the extent to which great achievements are dependent on carrying out the simplest kinds of tasks and responsibilities in an *exceptionally orderly fashion* is astonishing. Ivanka Trump (2009), who has started and managed several businesses large and small, appears to be a good example of excellent control over basics—consistently following an orderly, exacting, efficient methodology and achieving great success. As the great football coach Bill Parcells (1995) said, there is always a *way to win*; it just has to be found. That positive attitude and creative yet methodical approach to difficulties can become a habit.

Poise

It is helpful to nurture comfort and confidence in facing challenges, a sense of being centered and *poised* to cope, a certain smooth readiness for life and action. If an attitude of poise (alert preparation) is consciously developed, or simply taken on (as in self-hypnosis), then problems throughout the series of tasks that comprise a personal project can more easily be taken in stride. Poise is not only phenomenologically beneficial, it makes a positive impression on other people, as they observe someone who seems to have a situation well in hand. That impression also aids in success.

Heuristic Attitude

A good bet is usually better than a bad bet, but even good bets don't always come through. A fundamental aspect of ops thinking is that it is *experimental.* Situations change, people change, and what worked once may not work again. So, it is normally good to make an assessment, consider the probabilities, come up with a plan, and then, take a chance, to see how things go. But it is normally not so good to simply *assume* that some plan will work. A more self-protective attitude is to *hope* that it will work, that effort will pay off, but to be ready to *make adjustments based on results* in good Bayesian fashion.

Project Initiation

It might be said that the road to naptime is paved with good intentions. It is all too easy to come up with a good idea, and then delay it forever, all too easy to get stuck on *Someday Isle* (*someday I'll . . .*). Starting (initiating) and stopping (inhibiting) actions are fundamental to the nervous system, fundamental to computers, and fundamental to operations. Sheer recognition of the profound importance of action initiation is a good first step in mastering it. But there are other useful techniques.

Initiation Rituals

A project can be conceived of as having an effort curve, something like a learning curve. The slope is low at the bottom (where it is easier), then gets steeper (tougher) as the project progresses. So, it can be helpful to have some regular, fun, distinctive *starting activities* (an initiation ritual) that make beginning the work relatively easy and natural—such as putting on the coffee, turning on the light over the desk, and putting on favorite music.

Chunking and Milestones

The critical path usually can be broken down into chunks (sections) with milestones (clear finishing points at the end of each chunk). That bracketing of tasks helps a lot to make a project seem more manageable, makes attacking it easier, and makes celebrating completion more natural. Sometimes the order in which tasks need to go in is built-in, 1-2-3, but other times there is more leeway in which section could be tackled when.

Sequencing: Easiest First

It can be hard to know where to begin a project, and how. For people who have a little trouble getting going, often the best place to begin is the easiest step, or the step that looks like it will be the most interesting and fun (intrinsically rewarding).

Sequencing: Most Important First

For sterner types (like those who just missed the cut at the FBI or *made* the cut) the part to begin with is the most *important* part, even though those carry the most costs if they go wrong. Sometimes, jumping in and resolving one or more key issues will actually carry the day. Done. Move on. Therefore, the most important step might be the best one to start with.

Sequencing: Tight versus Loose

Realistically, tasks can get onerous, even if they lead to great rewards. Someone of bona fide stoic nature might want to put the nose to the grindstone and go until the job is done, however tough it gets, however long it takes. A *tight* sequence. No deviations, no irrelevancies.

But for regular mortals a *looser* schedule is usually more doable. A looser schedule means taking enough enjoyable breaks to make the job comfortable. That often includes regularly stopping to look over the work and provide some encouraging positive self-talk, like "I'm doing pretty well here" or "This is looking good," as well as taking a walk, getting a cup of coffee, and so on.

Cuing: Prompting

The sheer awareness that a *signal* can be paired with the beginning and ending of activity can help increase control over starting and stopping. For example, old-time factories used to have a whistle to start and stop the work day, bells often signal the beginning and end of spiritual services, and distinctive pieces of music accompany the beginning and end of a wedding, or accompany graduation exercises. Music can be fairly easily controlled; that stimulus then becomes a means of controlling behavior more difficult to carry out (a conditioned stimulus). For example, people frequently put on energetic music to help pump up enthusiasm for exercising.

The clock itself, and schedules, can help in starting and stopping. At a certain hour, or day, it is time to start; at a certain hour, or day, it is time to stop. Activities just need to be put on the schedule, and then the schedule needs to be checked regularly, and followed.

Physically, a particular area can be assigned as work, such as a certain desk for studying. Being at that particular place becomes a signal (conditioned stimulus) to start that job; leaving that place is a signal to stop that job.

Celebrations

Strictly speaking, behavior is a function of reinforcement, so it is crucial that an activity be regularly rewarded. That means a mini-celebration at the end of every milestone and a major celebration at the end of the project.

CLINICAL INDICATORS OF ACTION COMPONENT DIFFICULTY

A common complaint from anxious clients is that taking action is scary, from depressed clients that action is too draining, and from cautious clients that action is confusing, overwhelming, and risky. Just as an operation can be seen as the core of a system, action can be seen as the core of an operation—the actual sequence of motions along the path leading to the objective. So, it is no wonder that many of the best-laid therapeutic plans reach a stasis point when it is time to begin to move. Moving—actual action—is the biggest departure from what is sometimes called the "talking cure"; it requires the most effort, and it entails the highest risks. Difficulties are quite understandable; the situation is something like the shock of the new, or the alarm of the unknown.

OPS APPROACH TO STRENGTHENING ACTION: GENERAL

There are two basic ops strategies for enhancing *action capacity*. The first is to recognize this surprising truism: *the best way to build mental capacities is the hardest way—regularly working through difficult issues step-by-step* (White, 1991). As they say in Navy SEALS training, "The only easy day was yesterday" (Schoenberg, 2004). Serious life tangles usually result from trying to get around that essential building process (Berne, 1961). The second strategy is, in typical ops fashion, to strengthen the personality components *around* the Action component, creating a more capable action infrastructure through specific calibration.

OPS APPROACH TO STRENGTHENING ACTION: SPECIFIC CALIBRATION

- *Ideals*: Though the classical Greeks actually showed an amazing cultural balance between contemplation, action, and the artistic productions that often linked the two, their signature Homeric sagas blaze with the canonization of action (Bowra, 1957; Homer, 1963; Homer, 2004). Clients can enhance their inclination toward action by *adopting as a value the courageous engagement with reality*. They can enhance that value by reading materials like the Homeric sagas and by looking at movies and listening to songs that engender heroic feelings, by listening to motivational speakers, and by associating with confident role models, among other techniques.
- *Cognition*: The remarkable systems theoretician and Nobel laureate Herbert Simon in his book on administration (1976) emphasized the need to get as much sound information as possible, but also emphasized the importance of using that information as a *springboard to action* in a process he called *rational bounding*. Excellent information, and a great deal of it, is very useful. But at a certain point, action becomes necessary *within a frame of reasonable risk*. As indicated earlier, he termed this mentality *satisficing*,

by which he meant *accepting as a standard something reasonably satisfactory*, rather than waiting for the perfect moment to move, or insisting on the highest possible achievement as a minimum acceptable standard (both obsessive-compulsive tendencies). Satisficing encourages taking some kind of practical action, at some reasonable point in time, aimed at some reasonably satisfactory standard. Just so, the business world, as well as the military and sports, have a bias for action, an inclination toward relatively prompt, heuristic testing of ideas. Consciously absorbing this attitude of *pragmatic élan* can help encourage clients to shift from the considering phase (Reflection) to the doing phase (Execution).

- *Feelings*: Sights, sounds, and emotions are a little like the paddlewheel on a riverboat—without passion, not too much goes forward. However, *excessive* concern about danger and loss (anxiety and depression) are a little like paranoia—they can turn into a maelstrom if left to themselves. Likewise, *excessive* confidence, as with the *Titanic* racing through an ice field to disaster, can be dangerous. The basic ops strategy with regard to feelings and action is to learn to get excited about the opportunity to take action. That usually means pumping up the positive sensory and emotional aspects of action and outcomes, and learning to emotionally take in stride a degree of risk and even periodic failure. The character James Bond in the movie *Thunderball* at one point was unexpectedly, temporarily, completely trapped. When he grasped the situation, he shrugged off his anxiety and philosophically pointed out with that charming, reassuring, Ian Fleming–style, game-like attitude toward life: "You can't win them all" (Lycett, 1995; McClory & Young, 1965). (Of course, he never gave up.)
- *Decisions*: When clients have difficulty linking decisions and actions, it is usually because they feel a sense of panicky crisis, a certain *decision pressure*—the sense that a choice must be made quickly if disaster is to be avoided, or they feel *outcome alarm*—a fatalistic sense that once a choice is made, there is no going back, and the outcome may be disastrous. The basic ops strategy is for people to psychologically *elevate* themselves above the playing field of life, as it were. The Decision component can be conceptualized as an airfield control tower, sitting *above* and observing all the functional activity below, or as the sportscasters' box *above* the field—profoundly engaged with the process, yet slightly apart, and able to observe and think in a relatively cool and rational fashion. This paradoxical conceptualization of decision-making—involved yet objective—helps to free people from the experience of being trapped by circumstances.
- *Objectives*: As Stephen Covey suggested, "Begin with the end in mind" (2004, p. 95). Considering where everything is *going* has a marvelously narrowing effect on *how to get there*—on the range of action options. Often, uncertainty over action can be resolved by clarifying the destination.

The clearer the objective, the better. In operations, *focus is everything* (or a lot of it, anyway).

- *Communication*: Naval aviators (fighter pilots) are continuously embroiled in a very dangerous world of great prestige (Heatley, 1986). They *must* learn to adopt the attitude and corresponding *communication style* of what might be called the *hero's paradox*—verbally downplaying danger and being quite modest about success (almost a negative state), while radiating confidence and pleasure (a definite positive state). Part of the reason for this is that the pilots hear *themselves* talking that way. From their *own* communication they are reminded that the actions they take to survive and triumph are both manageable and satisfying—they convince themselves that it is not really too much of a big, scary enterprise, yet winning is very, very important. This *blend of modest yet confident communication based on focused, successful action* is part of the ops style, as presented in this book.

SUMMARY

If the core of living is operations, the core of operations is action. But for many clients, taking action is equivalent to fighting the fearful dragon on the way to the Golden Fleece. A great many techniques are available to assist in the initiation and continuation of action, from meditation and desensitization to task and time management. But perhaps the most useful technique is *embracing an ethos of action*, developing more of a self-image of being a person who approaches tasks with verve and completes them with dispatch. Something like that legendary adventurer Jack London—"A kind of genius reaching out toward the unknown" (London, 1989, p. 14).

THE ACTION-OBJECTIVE CONNECTION

Actions are almost a realm of their own. Intelligent people who have little trouble sailing through plays and poetry not infrequently come to grief, like ships on a hidden, rocky shore, when required to prioritize tasks or organize a desk! A key component that helps that process of separating the relevant from the irrelevant, establishing priorities, and lining up actions in a rational and effective order is having a vivid, precise idea of the objective and making sure that that objective is linked to the rest of the personality.

Chapter 10

Objectives

Personality Component 6

When Europeans finally, and proudly, arrived in the South Pacific hundreds of years ago in their bulky sailing vessels (Salmond, 2009), they were astonished to realize that the Polynesians, in their slim, relatively small, fast boats, with no modern (for that time) navigation instruments, seemed to have explored the entire Pacific Ocean, an area of millions and millions of square miles, from Hawaii to Easter Island to New Zealand, and had done so as far back as anyone could remember. As with most seemingly magical accomplishments, careful analysis began to suggest some answers.

The Polynesians thought, normally, not in terms of just going out on the ocean and sailing around to see what they might discover, but rather in terms of specific objectives. They consciously intended to go from Island A, so to speak, to Island B, in a relatively straight line. They even had stick charts, which had shells showing the location of islands (Mariner's Museum, n.d.), and they were highly aware of the water quality, currents, wildlife, winds, and stars that were the means to arriving at those destinations. Functioning in that world safely required amazing expertise and self-possession.

Of course, it is very difficult to connect something like extreme navigational competence in an island world with upbeat, alert, bold, joie de vivre personalities. But many people have commented over the years on the strikingly affectionate, aware, open-minded ebullience of the Polynesians (e.g., Bird, 1881/1998; Ellis, 1825/2004; Gauguin, 1994; Salmond, 2009), sometimes expressed as the *aloha spirit* (Say, 2004). And it may be plausible to suggest a connection between frequently and competently going long distances in small boats and having a relatively cheerful, adaptive personality with an ingrained sense of meaning, direction, and self-control.

At least, the Beat movement in America provides an interesting and possibly instructive contrast in traveling attitudes and traveler personalities. Jack Kerouac (1957/1991) wrote a famous, romantic, and somewhat beautiful book about traveling per se, with no real purpose in mind—*On the Road*. Kerouac knew he wanted to be a writer, but other than that he seemed to have little direction, or happiness, in his life (Johnson, 2012). He wrote, "I was just somebody else, some stranger, and my whole life was a haunted life, the life of a ghost" (Kerouac, 1957/1991, p. 17). Kerouac died, addicted to alcohol, at an early age. The signature Beat poem "Howl" contains the perhaps telling lines, "Whole intellects . . . who vanished into nowhere . . . who wandered around and around at midnight in the railroad yard wondering where to go" (Ginsberg, 1984). The prominent Beat writer William Burroughs was notorious for traveling around without much direction or purpose, typically addicted to substances, from New York to New Orleans, from Tangiers to Mexico (Miles, 2013). His writing has a strange atmosphere of desolate lostness—of thin, cold air; dusty corridors and muddy streets; and run-down buildings (Burroughs & Ginsberg, 1963). Perhaps not too surprisingly, Burroughs shot his wife to death in Mexico in 1951.

Life, as mentioned previously, is often described metaphorically as a *journey*, and sometimes as a *quest*. Thus, choosing a destination and traveling toward it can be considered an analogy to living. Choosing meaningful objectives and setting off toward them in a hopeful and perhaps even jaunty way almost defines mental health—the opposite of directionless or wasted, resentful or enraged. Setting objectives is even a core part of education and training (Mager, 1997). However, choosing meaningful objectives can be uncommonly difficult.

Objectives, as chief working variable in goals, have a tendency to define people and to define their lives. Therefore, it is not unusual for clients to resist choosing and committing to objectives. Often, they thereby hope to avoid feeling trapped in one identity, to avoid feeling tied down to one lifestyle. But that, of course, would be a false resolution to the stress of choosing. Without objectives, in principle, nothing happens. There is nowhere to *go*.

Another one of those paradoxes of living is that (1) *letting go of most options*, (2) *choosing a small handful* of really meaningful goals and roles, and (3) *developing expertise* that is satisfying and remunerative seem to be, overall, the most probable path to building a satisfying, secure existence (Bolles, 1982; Piskurich, 2002; Puppe, 1993). In the abstract formulation of the Santa Fe Institute, this is about *adaptive agents* finding a *niche* in which there are sufficient *supplies* (Waldrop, 1992). It is surprising, and perhaps disconcerting, how often clients' phenomenological difficulties in living partly can be traced back to a reluctance to settle on life objectives.

DEFINITION OF OBJECTIVES COMPONENT

Objectives, in the ops system, could be defined as *the particular target expected to satisfy a need or desire*. Objectives are premised on wants and probabilities, and are linked to outcomes, each of which typically involves some degree of success and failure. Some objectives, when successfully reached, result in satisfaction, while others don't quite end up as expected. And all objectives, in principle, when chosen and pursued result in all *other* objectives *not* being chosen or pursued, at least temporarily; these *opportunity costs* are for some clients stressful, and that discomfort sometimes explains a reluctance to choose objectives. And, of course, different objectives entail different degrees of effort and risk-taking.

Thus, objectives are a little more complicated than they look. However, to paraphrase a familiar motivational saying, *If you haven't decided where to go—how will you get there?* Establishing objectives is essential, and working through the complications is part of the assessment and commitment process.

RELEVANCE OF OBJECTIVES TO THE OTHER OPS COMPONENTS OF THE MIND

The mind is a system, and as such there is much back-and-forth, reciprocal interaction in something of a loose, dynamic arrangement. Each component is critical, providing a necessary function to the whole. However, in a way, the entire system *points to* the objectives. It even looks that way on the ops diagram of the mind. Reflection and Execution are largely *referenced* to the Objectives component.

HIGH-FUNCTIONING OBJECTIVES COMPONENT: OVERALL PORTRAIT

From an ops perspective, people with a smoothly functioning Objectives component generally sense that they know what they want; they experience the *confidence of focus*. They operate in a frame of action in which *purpose is paramount*. They can smoothly alternate from self-assessment of needs and desires on the left (in the ops model) to focused action on the right (in the ops model). They are comfortable with the process of translating ideas into action targets. They think strategically and tactically, so that they are focused on the larger picture and the relevant details, moving toward chosen targets in a determined way, adjusting as they go.

Not that high-functioning people can't have fun. For example, Queen Elizabeth I, who has a reasonable claim to being the founder of the British Empire, was highly focused on the survival, growth, and prosperity of Britain and was quite skilled in managing not only money and politics, but also the military, and famously defeated the Spanish Armada. However, she also loved pageantry, theater (she partially supported Shakespeare), music, dancing, dining, and witty conversation, and she not only made a point of enjoying herself, she encouraged her entire nation to do so. She was called by her people *Good Queen Bess.* She never married, in a shrewd political balancing act, but there is reason to believe that she was not absent high-status male attention (Axelrod, 2000).

HIGH-FUNCTIONING OBJECTIVES COMPONENT: SPECIFIC CHARACTERISTICS

Consciousness of Objectives

Perhaps the key characteristic of a high-functioning Objectives component is the *insight* that without objectives needs will not get met, and if needs do not get met, survival is threatened. The awareness that there must be objectives functions as a motivator.

Easy and Short-Term Objectives

Not all objectives have to be difficult and protracted, like getting a college degree. Often it is beneficial to establish quick objectives to test out ideas, or just to get an immediate reward. And sometimes it is OK to choose an objective just for the fun of it!

Hard and Long-Term Objectives

Unfortunately, a lot of the best rewards in life come from extensive focus and over a long period—playing Beethoven, learning golf, being able to do quadratic equations, earning diplomas, staying happily married, successfully raising children, paying off a car or house loan, building substantial investments, and acquiring pensions. Those seriously challenging objectives call for extra imagination and commitment, extra effort and patience, and that is often a clinical issue.

Balancing Objectives

Like a good investment portfolio, a well-run life tends to include both easy and hard, short- and long-term objectives, so that challenge is always there, but so are quick victories; an investment is always being made in the future, but there also is lots of low-hanging fruit to pick along the way. Thus, something good is always coming up sooner as well as later.

Specificity

A notorious bugaboo of setting objectives is fixing on ones so vague or ambiguous no one could tell if the objective had ever been reached, let alone how to reach it (Bossidy & Charan, 2002). By contrast, making X amount of dollars per year, or graduating from Y school, or marrying Z person are precise enough to be identifiable when the unknown is penciled in. Client issues often revolve around struggling with goals that have insufficiently specified objectives.

Tangibility

Words can seemingly convey an objective without quite referring to anything real. And, if exciting enough, language can carry people away without their quite intending to have it happen. *King Solomon's Mines, total victory in war, easy money, destiny and fortune,* and *unleashed potential,* for example, would seem to be at least highly questionable, if not more doubtful than that. *Unleashed potential* is particularly tempting—unhook the leash, and out charges a high IQ, a charismatic personality, or millions of dollars. (Even *possibility* is something of a mirage because almost *anything* is a possibility!) Such illusions of language and thought are sometimes termed *mystification* (e.g., Laing & Cooper, 1964; Laing, 1967; Marx, 1972).

Words can have a mesmerizing power, partly because the absolute flood of language to which people are exposed daily makes it necessary for people to adopt a working assumption that the words describing an objective are connected to something real, even though everyone knows that is not always true. Therefore, a certain mental discipline is required to *empirically verify,* or at least *empirically question,* at critical points, whether words indicating an objective actually refer to anything tangible (e.g., Ayer, 1936/1952).

Feasibility

As people become more conscious of the centrality of setting and pursuing meaningful objectives, they get better and better at hypothesizing what they are capable of, understanding what they are really interested in, and gauging how motivated they are. That means they get better and better at identifying objectives that, for them, are both appealing and feasible.

Visualization

A core operational skill is vividly imagining objectives. That assists in developing scenarios—virtual walk-throughs of an objective. Mentally seeing a target (such as a new job or a new house) makes thinking about it and then reaching it easier and more natural. Vivid *multimodal imagining*, including image, sight, sound, touch, smell, and taste, is a useful motivator (e.g., dinner in Paris, a house on the water, going through a graduation ceremony).

Quantification

If visualization, and multimodal imagining, give an overall sense of the objective, numbers provide the essential nitty-gritty. For example, a job opportunity may be glamorous, but a critical aspect of that objective usually is *salary*. A new house may be almost perfect in every way, but a critical aspect of that objective usually is *price*. "*How much*?" is usually a good question in the operational world.

Degree (and Range) of Demand

Objectives have a *difficulty level*—a degree of demand. The degree of demand is part of the *cost* of the objective. For example, completing medical school is more demanding than completing nursing school; becoming a lawyer is more demanding than becoming a paralegal. More prosaically, but perhaps more relevantly, putting in a new rose bush in the front yard is less demanding than putting in a tree. Estimating the demand and matching that to benefits and motivation is a critical part of the planning process.

How high should people aim? How low? The pediatrician and Olympic champion Stuart Walker had an interesting proposal (1986). He suggested that in choosing objectives it seems to work best to figure out the *range* in which success is likely. At the bottom of the range are the easy victories—sometimes useful, at least reassuring. At the top of the range are those demanding challenges that, with a lot of effort and a little luck, can be reached. That is what might be called the *challenge range*. The probability of succeeding should be *stimulating*, not intimidating or discouraging.

Outcomes/Results

As touched on earlier, there is a subtle linkage between objectives and what might *follow* the accomplishment of the objective—outcomes or, to use another common term, results. This is the concept of ROI (Phillips, 1997) and the concepts of MBO and MBR. In the ops system, that usually means that as a result of achieving the objective, the initial want was actually satisfied. That may seem obvious, or automatic, but it is not. Sometimes, a difficulty with objectives is assuming that a particular objective will satisfy a corresponding want, but then finding that it does not, being puzzled or frustrated. A dramatic example is fame, which often looks like a palace from the outside, but a prison from the inside. Therefore, a fundamental question in relation to an objective is, "*Will this actually satisfy me*?" Of course, the process of living is experimental, so it is often hard to tell in advance; nonetheless, it is a good question to keep in mind.

Wish versus Want

A surprising number of clinical conundrums revolve around the distinction between *wish* and *want*. A *want*, as previously indicated, is an actual survival-referenced need, or a *desire* that is not necessary, but still appealing. If a need is like a hat, to protect the wearer from the sun, the desire is like the feather on the hat. Or in an alternate metaphor—desire is the spice that enhances the meal. *Needs* and *desires* are combined in the term *want*. A *dream* is a major desire that might be very difficult or time-consuming to achieve. In the ops system, a want is something that, in reality, the client is ultimately, often after much discussion, willing to work for.

A *wish*, on the other hand, could be defined as: *an attractive option that, in reality, an individual is insufficiently motivated to pursue*. A wish is more like an emotional chimera or an intellectual mirage. By itself, unless the situation changes, there is perhaps little benefit to spending much time on pure, and perhaps tractionless, wishes. (Thus, *wishing* is different than *blue-skying*, which is a useful brainstorming technique.)

CLINICAL INDICATORS OF DIFFICULTIES WITH OBJECTIVES

Identifying and dealing with objectives is often stressful and confusing, when it comes right down to it, first, because the selection of objectives entrains so many other important variables. Clients with issues with objectives may say that they are unsure of themselves and, therefore, unsure of what objective to aim at in life. The may complain of feeling pressured by friends, family, and society to choose goals that don't seem to fit their personalities. They may have ended up with careers or partners that are a strain or mismatch for what later seem to be their true dreams, their true selves. There is often an atmosphere of restless uncertainty, of casting about for a true destination.

A second common difficulty is that clients become *habituated* to the objectives they have selected. That means that, after a while, the objectives—like house, spouse, children, and job—seem boring and unsatisfying. Simultaneously, people become *sensitized* to all the things they *don't* have—like, say, a condo in Monte Carlo, a Ferrari, or a bestseller.

OPS APPROACH TO STRENGTHENING OBJECTIVES: GENERAL

There are, first, several general approaches to dealing with that uncertainty about objectives. Discussing in a comfortable, supportive, and open-ended way what might be appealing objectives can help clarify things, using active listening. Free association, journaling, and various forms of artistic or poetic work can point clients toward their true desires. Various behavioral methods are available to ease that identification and assessment process, for example, the roster of classic stress reduction techniques, such as desensitization, presented in Davis, Eshelman, and McKay's excellent and thorough book (1982).

Second, there are two general clinical approaches to the *habituation-sensitization problem*. The first is to *refresh* awareness of the importance of those selected objectives, that chosen lifestyle, something like the belated recognition by the hero of *It's a Wonderful Life* (Capra, 1946) as to what a wonderful life, family, and town he actually did have. The second general approach is to take the discontent seriously, at least heuristically, and to assess the current objectives being pursued, and perhaps brainstorm new ones. Often that procedure results in a tonic awareness that dropping old commitments and establishing new ones can entail a lot of complications that don't look quite so appealing on closer inspection. And the reality is that any new objectives will also be subject to the process of habituation down the road. As the psychiatrist Martin Blinder pointed out (1989), it is remarkable how often a second spouse resembles the first and how much a second marriage resembles the first. Then again, new objectives and a new life may start to look like the best way to go.

OPS APPROACH TO STRENGTHENING OBJECTIVES: SPECIFIC CALIBRATION

- *Ideals*: Granted, there is something beautiful about absolutely pure existence, like Henry David Thoreau (2008), retiring in the mid-nineteenth century to a simple life in the woods at Walden Pond to contemplate the mysteries of nature and life. And certainly there are moments where doing absolutely nothing, almost *disengaging* from the stream of events, is blissful and revitalizing—like listening to peaceful music, doing yoga, going for a walk at dawn, or sitting down to a delicious meal in the evening breeze and watching the sun go down. There is something *ideal* about that kind of experience, something intrinsically *valuable*, something almost *transcendental.* Indeed, there was a whole movement in America in Thoreau's time called *Transcendentalism* (Emerson, n.d.). Nonetheless, it is probably not enough. Thoreau lived like that for only two years, and he had to get someone else's permission to do it because he didn't own the land around Walden Pond—it was owned by Ralph Waldo Emerson, who was a little more focused than Thoreau. So, the first step in developing a strong Objectives component to the mind is to deeply appreciate experiences for their own sake, but also to recognize the power and necessity of having pragmatic objectives in the management of a successful life.

- *Cognition*: Objectives are rarely black and white. They usually are a combination of characteristics. Those characteristics can be rated from one to ten. The top few traits and the bottom few traits together form a *profile* of the objective. For example, if the top two traits are rated at ten, and the bottom two traits at one, that objective receives a score of a little over five, resulting in a middling option. Profiling sounds complicated, but it is actually fairly natural. Looking at a house, for example, a person might say, "The bedrooms and bathrooms are great, but the yard and garage need a lot of work." Or, in considering a potential partner, a person might say, "Great smile, great sense of humor, but not so reliable." Conscious profiling is simply more exact, and thus typically more effective.
- *Feelings:* It is easy to get all pumped up or all discouraged over objectives. They often seem fabulous but out of reach, or well in hand but boring. There is a certain art to getting excited about and staying excited about feasible objectives. And there is a certain art to maintaining a poised emotional distance from objectives that could go right or go wrong. In a high-functioning ops system, feelings enthusiastically engage with objectives without losing perspective.
- *Decisions:* "Follow the money" is a common saying. Perhaps it could be modified to "Follow the money to the objective." The *why* behind the money is the real issue. Similarly, in the mind, and in a life, once the objective (the immediate *why*) is selected, almost everything else follows that. The objective in effect *commandeers resources.* Which can be alarming. A high-functioning Decisions component is shrewd and honest in assessing needs and desires, probabilities and motivations, and then in selecting an objective likely to be interesting and effective without getting unduly tangled up in either worries of failure or fantasies of success. So, that informed and balanced, reflective yet bold style of selecting objectives needs to be consciously and continuously refined.

- *Action*: Within the ops system, *means and ends* could be essentially considered *action and objectives*, which form a sort of *golden link* within the mind—this is where the work really gets done. A high-functioning Action component is as carefully oriented toward objectives as a pilot focusing on a landing strip or an executive officer on a ship listening to a captain. It is useful for people to learn that knack of throwing themselves into projects once the objectives have been determined in a behavioral flow.
- *Communication*: The basic approach to being positive in communicating about objectives, once they have been determined and committed to, is to *focus on the benefits of winning and the costs of failing*. This is the basic *motivational frame of mind*. Simultaneously, and paradoxically, it is good to keep a weather eye on whether the objective looks like it *is* actually being achieved or not. Sometimes, despite the best of intentions and efforts, the only honest and sensible thing to say is, "It didn't work."

SUMMARY

In a way, the objectives selected define personalities and define lives. How can something be achieved if it does not become an objective? And if nothing is achieved, what then? Objectives are so important that they can distort the goal-seeking process, either by excessive and unrealistic ambition, or by evading the process of selecting and committing to them. The art of choosing objectives is to choose ones that are challenging, exciting, and useful, while not being too risky, draining, or discouraging.

THE OBJECTIVES-COMMUNICATION CONNECTION

It may not be quite as bad as Nietzsche suggested in his discussion of good and evil (1886/1973), with the conscientious and typically not-so-articulate workers toiling away toward their objectives being the slaves and the smooth, charismatic, well-spoken elites signing orders and making speeches on the way to *their* objectives being the masters. However, almost undeniably, persuasive terms, tight logic, compelling stories, and bracing repartee are aces in the game of life. Effective communication per se wins a surprising number of conflicts, negotiations, and fortunes. (At the very least, it is good for style points.)

Chapter 11

Communication

Personality Component 7

Is the television comedy series *Seinfeld* really about nothing (Barron & Wolff, 1989)? Is it not about the failure to communicate? In relation to communication, Jerry can be seen as qualitatively different from the other three—George, Elaine, and to some extent Kramer. Jerry is connected to social reality. He effortlessly grasps an effective *theory of mind*—he understands that other people have feelings, that they have goals and lives, and that dealing with other people successfully means listening to them and fitting in with their programs in a reasonably sensitive, tactful, competent way. Almost all of that, of course, escapes his three comedic friends. Over and over they fail to register and accommodate adequately to other people's communications. They never learn, somehow. Which is funny, but not a blueprint for success—none of the three of them has a real career or even a steady job most of the time, they seem to have no other friends and almost no steady partner, and they are usually puzzled, upset, and then derailed by some piece of utterly ordinary social interaction.

Communication in some form has doubtless been around as long as humanity has. Even Cleopatra had a catchy way with words—at the end of an extensive legal-business document, she wrote simply *ginesthoi*, which potently means *make it happen* (National Geographic Society, 2010). But for a long time, perhaps until the work in the nineteenth century of linguists such as Ferdinand de Saussure (De George & De George, 1972), it seems like it was just something people did in and around the process of living. *Communication* per se may not have assumed its perceived central role in human functioning until Norbert Wiener (1961) developed the idea of *cybernetics*. In the cybernetic formulation a communication loop runs from desire for a particular pattern of behavior to evaluation of the pattern. If feedback information indicates that the actual pattern of behavior varies from the desired pattern of behavior, then new input brings the actual behavior in line with the desired behavior. In a machine, this is a fairly direct control loop. This is the way a thermostat communicates to a heating-and-cooling system that the temperature is too high or too low; then the system pumps out warm or cool air to bring the temperature within the desired range. In a simple system, the communication loop governs system behavior.

As alluded to earlier, the anthropologist Gregory Bateson and the psychiatrist Jurgen Ruesch (1951) then took that communications idea and applied it to human life difficulties. The hope, of course, was that in some directly causal, instrumental way, changing the communication would change the experience of living, change character and ideals, change action and outcomes. It hasn't quite lived up to that early enthusiastic press, perhaps because of the difference between machine causal responses and human decision-making (incoming information doesn't seem to directly govern human decisions, unlike a simple servo-mechanism). But Bateson and Ruesch did succeed in showcasing communication as a central concept in understanding human functioning and relationships.

Profoundly poor communication is rare and can be amusing, as in the case of most of the main cast in *Seinfeld*, or the movie character Bridget Jones, who was an "appallingly bad public speaker" according to her eventual boyfriend, Mark Darcy (Bevan, Cavendish, & Maguire, 2001). But *moderately* problematic communication is not uncommon and is usually unnerving, as attested to by the average individual's terror of giving a speech, or even meeting a seriously important person. Even the celebrated French actress Brigitte Bardot at the height of her fame was nonplussed as to what to say to the Queen of England when she met her (Bardot, 1996.) Dale Carnegie (1981) made an entire career out of teaching people how to improve their communication; Carnegie training programs are still available and may be a useful option for some clients.

DEFINITION OF COMMUNICATION COMPONENT

As mentioned previously, in the ops system, communication has two primary aspects—feedforward and feedback. *Feedforward* could be defined as *attitudes and statements intended to support the pursuit of objectives*. *Feedback* could be defined as the *candid and balanced registration of incoming information, especially about degrees of success in the pursuit of objectives*. (This is a phenomenological definition—feedback consists only of what is *received*, not everything that was *sent*.) But those communication aspects are really a little broader and more intentional than that. Feedforward is, overall, positive communication aimed at trying to make life a success. Feedback is, overall, an attitude of poise and frankness in surveying the landscape, and in seeing whether sufficient success is occurring, and then a willingness to take corrective steps if things are out of line. Thus, dealing with negative feedback, which is not infrequent in life, partly because it has a corrective, navigational function, also is a part of being successful. Feedforward (upbeat proposals) and feedback (honest evaluations) work together; at best, they synergistically interact in building success in life.

RELEVANCE OF COMMUNICATION TO THE REST OF THE OPS SYSTEM

The Communication component in the ops model of mind is conceived of as the primary *interactive interface* between the individual and the surrounding world. This is where individuals get to present their case to the world (feedforward), and this is where the world gets to give its reactions to the individuals (feedback). The personality components *align* to generate effective, useful projects and *cooperate* in adapting to incoming information. That internal alignment and cooperation and that adaptation to external performance information form the basic process of survival.

HIGH-FUNCTIONING COMMUNICATION COMPONENT: OVERALL PORTRAIT

In the ops paradigm, people who communicate well speak and write confidently, clearly, and persuasively. They are enjoyable company and are sought out by others, interesting as well as courteous in a natural way. They are alert to life's events and personages, and enjoy sharing bits of news and their thoughts and reactions to the day's occurrences. They are easily met—friendly without being overbearing or intrusive, sensitive without being anxious, respectful without being intimidated. They accept that the perfect and the real only approximately overlap, bestowing on human life a certain fundamental absurdity, which they find somewhat amusing but, if tragic or in need of correction, a reason to try to improve things. They laugh easily, but never cruelly, and enjoy sharing jokes and good times, while maintaining an underlying, responsible seriousness. They are well able to advocate for themselves and their projects without having to resort to deception or fabrication. Thus, they believe in their mission and are comfortable and skilled at emphasizing, in a calm but passionate way, the virtues of their plans and ambitions.

Effective ops communicators are also good listeners, attentive and accurate in receiving information. While dedicated and emotionally involved in projects and outcomes, they are able to maintain that sage and sensible distance from information about progress and outcomes. They are able to maintain emotional balance and a steady forward pace despite periodic confusion and disappointments. They are able to hear what the world says about their performance; outright defenses marginally distort the validity of their perceptions. Rather, they are inclined to seek safety in a more functional way—increased effort, greater flexibility, more creativity, consultation with others, a philosophical perspective, or a sense of humor.

HIGH-FUNCTIONING COMMUNICATION COMPONENT: SPECIFIC CHARACTERISTICS

It may be useful to highlight prominent aspects and examples of strong communication. These qualities and role models then can be cherry-picked according to the preferences of the client and assembled into a personal template of high-functioning communication.

Radiating a Sense of Well-Being

President John Kennedy was known as a captivating and powerful writer and speaker. Many people are aware of his electrifying inaugural speech in which he famously told the public to ask not what the government could do for the public, but rather to ask what the public could do for the government. It was profoundly well put, but part of what made it convincing is that he himself was such a charming personage who radiated a charismatic sense of cheerful security (Hamilton, 1992; Wills, 2009). Similarly, speakers from the National Speakers Association (NSA) typically radiate a profound sense of grace and élan, as do many entertainment personnel. NSA and acting coaches can be a resource for clients.

Strong Vocabulary

The classic key indicator of intelligence and thus, to a degree, of wisdom is having a broad vocabulary. If there is one quick test that will be given to a client to estimate intelligence, it is a vocabulary test. Having access to a broad range of terms makes it easier to select exactly the right term as well as to understand what is being said by others, no matter how sophisticated. Lists of vocabulary items are readily available for memorization.

Simple, Clear Language

Listeners usually find simple language more convincing than complex language, since weak ideas can be obscured in convoluted statements, and people are generally aware of that. There is a certain art to speaking and writing in crisp, short sentences that clearly yet vividly convey ideas, and tying them to verifiable grounds in a natural way. The novelist Hemingway (1964; Phillips, 1984) and the language professors Strunk and White (1979) are classic models of crisp, concrete communication. Reading and imitating such models can help clients improve the directness and clarity of their language. Correspondingly, *proactive listeners*, when appropriate, request that speakers or writers communicate in simple, verifiable terms; they are not interested in unnecessarily convoluted or untethered communication.

Convincing Advocacy

Lawyers and other advocates (e.g., debaters) often generate a neutral, in-house *brief* that sums up a situation (Calleros, 2014; Ericson & Murphy, 1987). That synopsis captures the basic players, their intents, key actions, the main background, and probable outcomes in a way that is balanced and well grounded. *Then* the idea is to slant the description of the situation in a way favorable to the speaker or writer. It is not all that different from a parent *selling* a child on taking vitamins, eating a healthy meal, or going out to play in the sunshine by emphasizing the aspects of those options that work for the child. Also, the Timberlawn family systems research (Beavers, 1977) showed that parents of happy children explicitly teach their children to be able to effectively stand up for themselves rhetorically. A strong Communication component can advocate confidently and effectively for the self.

Personal-Public Balance

At the same time, part of what makes someone a convincing communicator is the impression listeners get that the speaker or writer is not purely self-interested, but that the speaker or writer is also concerned about the listeners in particular and humanity in general, which is to say, appears to exhibit fairness (Ericson & Murphy, 1987). Therefore, an effective communicator takes into account the needs and desires of others in making a pitch and engages in empathic listening—this shows a concern that comes across clearly to others.

Pragmatic Focus

Similarly, an effective Communication component is aware of the need to, in some way, have communication usefully *deliver* to the listener or reader.

Audiences may want to be entertained, but they do expect to get something useful in exchange for their attention (and often for their fee). A utilitarian self-check for the communicator is, "How is what I am communicating going to be helpful to my audience in a practical way?" It is easy, without reminders, to get off-track, and thereby lose credibility with listeners who are looking for practical answers to their concerns.

A Sense of Humor

Everyone appreciates someone who can help them laugh at the "the slings and arrows of outrageous fortune," as Shakespeare put it. As indicated earlier, helping people laugh essentially comes from *cultivating a humorous slant* on life—recognizing that life is a little absurd (because it frequently does not quite fit or work out right), wisely accepting that fact, and developing the knack of making those relatively harmless absurdities amusing rather than just irritating. Jerry Seinfeld himself can be seen as something of a comedy therapist along those lines—comfortably seeing the humor in people's miscalibrations with life. High-functioning communicators are often a source of pleasant banter, as humor is a relatively mature way of coping (Vaillant, 1977). A little humor *humanizes* a speaker.

CLINICAL INDICATORS OF DIFFICULTIES WITH COMMUNICATION

Clients tend to have two sorts of complaints in the area of communication. First, they experience difficulty in either saying anything at all or in saying the right thing (regarding feedforward). Second, they often find themselves relying on defenses more than they would like, to the extent that they are aware of them, blocking out important but unpleasant incoming information (regarding feedback).

More generally, communication is a process too often fraught with conflict and uncertainty. It doesn't seem as smooth, comfortable, or effective as clients would like. It doesn't seem like it takes much to get derailed from a productive conversation into boring, threatening, or unfamiliar topics that are anxiety-provoking or discouraging. Many clinical discussions revolve around conversations from daily life that are repeated in session and then analyzed, broken down into conflicting agendas and often angry words. Then there are often reenactments, role-plays of how things might have gone better. For many clients, improving communication is critical to living better.

OPS APPROACH TO STRENGTHENING COMMUNICATION COMPONENT: GENERAL—FEEDFORWARD

There are numerous ways to develop the ability to present ideas convincingly, appealingly, as partly alluded to above. Reading widely helps; good writing shows how to structure and express ideas. Getting involved in discussions of the relative merits of sports teams or entertainers can teach how to argue a point while maintaining cheerful composure. Schools usually have debate teams that can help polish analysis and presentation. Toastmaster's offers excellent basic speech training, and NSA shows how to polish skills and attitudes to a high gloss. Studying great speeches can help—Kennedy and Churchill have been recorded, and thus can be listened to and imitated. Even television, the movies, and interviews with movie stars can provide useful role models for appealing speech. The Internet is a cornucopia of useful options. There are many business writing classes that teach focused, concise, persuasive communication. Comedy shows, books, and movies can help teach actual jokes as well as comedic timing and intonation, and practice can help develop tasteful, pleasing repartee.

OPS APPROACH TO STRENGTHENING COMMUNICATION COMPONENT: GENERAL—FEEDBACK

Feedback usually comes in two forms—written and, more commonly, oral. Written feedback might occur in grades for a student or a performance evaluation for a worker. Oral feedback is ongoing for most people in the normal course of living. The most important part of dealing with feedback is simply *registering* the information; the psychoanalyst Karen Horney (1950) pointed out that people can listen and even repeat back what they have been told without quite even having the information *impact* them or, as she put it, *register*.

If the feedback is negative, that could be painful. Then the mind can distort the information in some typical defensive ways—denial, minimization, projection, and repression, for example (e.g., Vaillant, 1977). A high-functioning Communication component needs to learn to listen to negative feedback and learn from it without excessive self-protective prejudice. Of course, developing that knack is rather challenging.

Learning to hear feedback about performance, appearance, intelligence, talent, and so on in a calm and thoughtful way takes conscious effort. Perhaps the first step is learning to take a deep breath and think *about* the feedback. It could be accurate. It could be inaccurate. If it comes from a strong source, it might well be profoundly on target and worth considering. If it comes from a weak source, it might well be highly misleading and not worth taking seriously.

An essential aspect to the idea of looking at living within the ops systems is that actions have been *hypothesized* to be effective; thus, the dominant attitude should be one of cool appraisal. Success suggests that the initial judgment and planning process were good; failure suggests that perhaps another approach would work better. With this *experimental* attitude toward people living and working together, feedback is much easier to hear and deal with constructively. That *heuristic attitude toward success and failure* can be consciously developed through prompting self-talk and self-reinforcement. Therapists can encourage in clients a learning approach to feedback through modeling objective curiosity toward feedback, including that from the client regarding the nature and progress of therapy.

Jerry Seinfeld (2015), in one of his droll remarks, has shown how confusing feedback can get by pointing out that many marital problems revolve around a wife saying to a husband that he doesn't listen to her. "I have never heard my wife say I don't listen to her," he mentioned reassuringly. But then he added, less reassuringly, "She *may* have . . ."

OPS APPROACH TO STRENGTHENING COMMUNICATION: SPECIFIC CALIBRATION

- *Ideals*: If self-development is taken seriously, and cool, smooth communication attitudes and skills are actually brought to a sharp edge, that results in *power*. And power can be misused, as noted in, for example, Ann Ruth Willner's review of impressive but deceptive political leadership—*Spellbinders* (1984). Therefore, it is important that communication competence be developed within a *larger frame of compassion and justice*. That constraint may be experienced as a drag on motivation and excitement. But in the long run, *persuasion framed within wisdom* is not only more satisfactory to listeners, but also more satisfactory to the self, on those quiet autumn afternoons looking back down the road along which one has journeyed. Once again, this is a character-development issue.
- *Cognition*: As the consultants Larry Bossidy and Ram Charan (2002) pointed out, clarifying the exact means that will lead to specific ends can be very tricky, exacting, and exhausting. But, sometimes, there is no other way. It is the same with communication. Learning to communicate *convincingly, factually,* and *logically* as well as *attractively* can take some doing. Numerous sources are available for assistance, such as Hollywood coaches like Michael Hauge and Christopher Vogler (2003), political role models, great plays, and logic books as well as therapists. Similar resources are Lilyan Wilder's *Professionally Speaking* (1986) or Patricia Fripp's *Presentation*

Skills Hollywood Style (2011). But there is really no way around the hard, slow work of polishing *sound reasoning linked to charismatic expression*. Understanding that helps to motivate the development process.

- *Feelings:* Of course, it is not *all* work! There is a certain thrill to learning to speak with more panache, to debate with more precision, to negotiate with more verve and composure. And in personal relationships, it is good to learn to talk and listen with more depth, more affection, and more calmness linked to enthusiasm and thereby to see relationships deepen and to become more interesting and satisfying.
- *Decisions:* This is the pivot point for communications. The Reflection side asks, "What is the right thing to say here, what is the most accurate thing, and how does it feel to consider sharing that?" Then the Execution side asks, "What is the objective with this audience, considering its feelings and needs, and then what are the spoken or written words that would work for that purpose?" The overall question is, "*How can I be both truthful and effective*?" All that, as most people know, is actually a fairly complex question. No wonder that people often consult with others, such as partner, family, best friends, or even a board of directors, and over a period of time, before making a major announcement. No wonder that thinkers from Lao Tzu (1963) to Marcus Aurelius (2006) have often counseled silence and meditation in place of words. This is similar to what Ernest Hemingway said, "The hardest thing in the world is to write straight, honest prose" (Wilson, 2009, p. 11).
- *Action:* The usual issue connecting communication and action is *how* to express something. And that depends on the audience. The business world has a 360-degree communication process that offers a useful guideline (Gebelein et al., 2004). The *360 degrees* refers to communicating with supervisors, peers, and supervisees. It is surprising how many interactions fit into that *up-middle-down* framework. Communication with people one tier up (e.g., an actual supervisor, a judge, a customer, or an audience of a talk) should be frank, straightforward, thorough, concise, factual, and overtly respectful; feedback *from* an upper tier is something like an *order*. Communication with peers can be more chummy and indirect, more oriented toward joking and fun as work gets done; feedback can be discussed more readily and is more susceptible to evaluation and counter-feedback. Communication with subordinates (e.g., actual employees, or valet and wait staff), though pleasant and thoughtful, is often more brief, to the point, confident, and authoritative. All of that really applies to nonwork communication as well (e.g., talking to a grandmother, versus a partner or friend, versus a child). The psychiatrist Eric Berne (1961) expressed a similar idea in a transactional system—Parent, Adult, and Child. The style and content of communication should fit the audience.
- *Objectives:* Mark Twain (1995) once was asked to give a talk to a children's group and to give them some good advice. He admonished them to

obey their parents *when their parents were around.* He encouraged them to be respectful of their superiors *if they had any.* And he reminded them to be careful about lying, because *it was easy to make a mistake and get caught.* Of course, that was all in fun, and the objective of this professional humorist was to have people laugh. It is, by contrast, surprisingly easy in communication to just launch intuitively into statements and forget the purpose of the communication. As the famed sports agent Mark McCormack pointed out (1986), a good way to avoid self-destructive remarks is to consistently ask, essentially, "*What do I want to have happen as a result of this communication*?" (This could be thought of as *CBR—communication by results.*) Thinking in terms of objectives and outcomes can tremendously focus communication, as well as improve the likelihood of the communication being successful, and avoid getting lost *inside* the statements. Staying *outside* the statements and holding as primary the *purpose* of the communication takes intention and practice, but it is a skill that can be learned.

SUMMARY

Communication is the interactive, persuasive link between the interior realm and the exterior realm. Communication is the set of messages that tell the world what the individual is up to, and would like, and it is the set of messages from the world as to how it sees the individual and what it will provide. Communication is essential in the adaptation process. Since so much of human life is conducted symbolically, and since all those symbols don't always mean what they sound like and look like they mean, a strong and capable Communication component consistently assesses language and images for *validity*. As stated by the prominent cognitive linguist Ray Jackendoff (2012, 245), "Sentences are true or false on the basis of how they correspond to the world."

THE OPERATIONS MODEL–PSYCHOTHERAPY CONNECTION

The operations model of personality comes to seem natural—there are ideals and values guiding the personality, there are cognitions and feelings providing information for decisions, there are actions aimed at relevant objectives, and there is communication about degrees of success. Nonetheless, it may not be obvious how all that fits in with a client's life or how to deal with clients so they can meet needs and pursue dreams. What remains to be clarified is the psychotherapeutic, *operational method* of helping clients deal with difficulties and challenges in life.

Part III

CLINICAL APPLICATIONS

Chapter 12

Anger

Defending Territory

Madness. Hardly anyone uses that term anymore. Perhaps it got too popular, as a term or as a state of mind. As it lost its distinctiveness, it lost its meaning. It used to have a clinical import, implying loss of reality contact, or *psychosis*. It also had a more fashionable air, meaning something like *wild abandon*.

The long and storied history of the word *madness* and of mental illness generally (Caplan, 1969) primarily involves what might be thought of as *loss of operational competence*—the ability to meet needs, carry out rational actions, communicate coherently and meaningfully, and cooperate with others. But the interesting part is that the root—*mad*—is a synonym for *anger*.

That etymology suggests that the traditional commonsense interpretation of mental illness is *anger out of control*. And anger does seem to be broadly involved in personal difficulties as well as various kinds of injury. It frequently seems to be involved in defiance, threats, physical conflicts, marital conflicts, family abuse, and addictions, not to mention political, doctrinal, legal, or military conflicts. Rage often seems to be a part of decompensation and hospitalization. Depression is sometimes defined as *anger turned inward*.

Ironically, anger per se appears nowhere in the international diagnostic manual (World Health Organization, 1992), although a research coordinator for the DSM once stated that anger was being considered for a place in the diagnostic taxonomy (Thomas Widiger, personal communication, 1994). Anger appears to be something of an orphan category with respect to formal diagnostic systems and may be one of the few issues that has a treatment assigned to it—anger management—in the absence of a formal label.

Perhaps the idea of giving anger a bye started with Aristotle (1961). Though Aristotle might be considered ground zero for crisp, categorical thinking, he felt that upsetness and tragedy observed, as in a play on stage, had a cathartic effect on people—"Tragedy, then . . . through pity and fear effecting the proper purgation of these emotions" (Aristotle, 1961, p. 61). In other words, anger and violence were good things, on stage, at any rate.

Gestalt therapists from the 1940s and 1950s (Perls, Hefferline, & Goodman, 1951) were scandalized at the anxious, rigid, excessive self-control they perceived in the mid-century American public. Gestalt therapy can be seen as an arena for the relatively unfiltered sharing of attitudes and emotions, including anger (Shostrom, 1968). Therapists, in discussing cases, often mention the value of letting clients *ventilate*. There may be in all that, to some extent, a ghost of a presupposed biomechanical theory in which anger is seen as something like steam in steam engine, or poison from a snake bite, and bleeding off the poison or the excess steam is healthy for the system. The poet Dylan Thomas may have given anger a sheen of glamour in his valiant lines "Do not go gentle into that good night/Rage, rage against the dying of the light" (Thomas, 1952/1998, p. 252). The television sitcom character Archie Bunker and his angry remarks were conceived by his entertainment creator as cathartic and as resulting in a healthful clearing of the psychosocial air (Lear, 2014). News media often seem to present conflict and anger, battles and threats between individuals and groups on the public stage as if those were good things or, at least, electrifying and fascinating, and often appear to rearrange community needs, programs, and personnel (i.e., reasonably sensible, predictable, and manageable operations) into raging, disorienting dramas (e.g., Mazzeo, 2012).

However, the classic work *Principles of Behavior Modification* by the Stanford University psychologist Albert Bandura (1969) suggests that though there may be some benefit to people "getting something out of their system," as the expression goes, that benefit likely is overwhelmed by the costs of exposure to negative models, because modeling is powerfully associated with *imitation*; thus, *negative models are associated with negative behavior*, and that would include anger and violence. Therefore, it may be wise to consider the issue of anger anew and to consider how anger—which seems like a fairly common and natural emotion—can be integrated into a high-functioning personality.

DEFINITION OF ANGER

Anger could be defined within the ops system as: *an aggressive, ultimately protective reaction to a perceived threat*. It involves multiple aspects of the personality as well as physiology and is often involved in social conflict. At its best—spirited self-protection—it assists in regulating the inappropriate intentions and actions of others.

THE FUNCTION OF ANGER—AN OPS PERSPECTIVE

Human beings are *territorial* (Ardrey, 1967). For example, if a person has an apple orchard, and that person's family depends on that orchard to survive, then interlopers stealing apples is no small matter. How to protect the orchard? Anger is often a starting point.

To be angry is to be *aroused*—heart pumping, face turning red, eyes narrowing and flickering, the voice becoming course and loud, shoulders hunching, and finally, the arms and hands, perhaps holding weapons, may be waving in a counter-threat—this is a familiar display. All of this not only functions as a signal to others—a warning—but serves to pump up the individual in preparation for a possible protective fight. In addition, as many people have noted, the surge of adrenaline engenders a devil-may-care, bet-the-farm, take-no-prisoners, slightly drugged mentality that favors the defender (and may incline casual interlopers toward the speculation that an easier orchard to steal from might be just down the road).

There seems to be, perhaps strangely, a certain social wisdom in anger, in true systems fashion, a *reciprocal logic*. People develop systems that provide them with satisfactions, the interpersonal equivalent of orchards. Other people don't always know exactly where the boundaries of those systems are or whether those boundaries absolutely *have* to be respected. Anger tells them.

For example, most people draw a subliminal boundary around their workspace and feel threatened and therefore irritated if someone invades that space without permission. For that irritation to be effective as a *signal*, other people need to be *sensitive* to it. There has to be some kind of unspoken match between the person getting *angry* and the person who is the target of the anger feeling at least a little *alarmed.* Thus, if Person A needs to find a report in Person B's work area, and Person A is found rummaging around without permission by Person B in Person B's work area, there is a certain ritual both go through—Person B calling Person A to account, Person A feeling alarmed and apologizing to Person B, and giving an immediate and clear explanation as to the reason for the boundary violation, then Person B accepting the apology, and offering permission and assistance to Person A. The semi-anthropological ritual verifies the mutual endorsement of the unspoken territorial contract.

People who claim more territory than they have quite earned (e.g., narcissistic tendencies), or who have very low sensitivity and care little if they cross boundaries or make people angry (e.g., antisocial tendencies), throw off this somewhat delicate *interpersonal arrangement of rights and responsibilities.* People who blithely ignore protective anger actually run a risk, in that they obligate the defender of boundaries and territory to escalate toward the use of force, such as legal action or violence. Thus, coming to a rational understanding of anger, and acquiring sophistication and skill in both expressing and dealing with anger, might be of general psychosocial benefit.

TYPICAL CLINICAL COMPLAINTS REGARDING ANGER

Clients with anger issues often don't present themselves for care. Anger is often accompanied by an experiential rush not unlike mania, and is thus frequently egosyntonic (i.e., the issue is experienced as familiar and comfortable). In addition, anger sometimes provides interpersonal leverage—power in relation to others, such as a spouse or employees—that people may be disinclined to relinquish. Furthermore, anger not infrequently provides a certain counterphobic protection against fear, especially in relation to people who abuse others and who have been abused themselves. Therefore, courts and human resources (HR) departments are often the impetus behind clinical contact regarding anger.

Nevertheless, when clients do voluntarily present for care regarding persistent excessive anger issues, they tend to present similarly—they feel trapped by their anger. Just when they were negotiating amicably with someone, or just when they were on their way to a promotion, or just when they met a nice person who might end up being a partner, their excessive anger slipped out, almost like a Freudian slip, betraying a deep vein of rage and destroying their prospects. And that happens to them again and again. Since most people don't respond well to marked, inappropriate anger, it becomes something of a social stigma, a spoiled identity (Goffman, 1986), which can be difficult to repair. Thus, anger is a genuine clinical concern.

PORTRAIT OF HIGH-FUNCTIONING OPS PERSONALITY IN RELATION TO ANGER

The actress Audrey Hepburn has a reputation for having been an exceptionally nice person, affectionate, but always under control, always focused and functional. However, there is a semi-legendary story about Audrey Hepburn and anger (Ferrer, 2003). The elegant, evocative song "Moon River" had been written for Audrey for the movie *Breakfast at Tiffany's*. At a meeting that included Audrey, the head of the studio unexpectedly announced that, without question, the song "Moon River" was going to be eliminated from the movie. Hepburn, instantly furious, shot out of her chair and flatly defied the studio executive (for whom she worked) and his plan. She got her way, of course—she protected her territory—and the song became a movie favorite.

Some classic actors, such as Humphrey Bogart (Wallis & Curtiz, 1942), Michael Douglas (Douglas & Zemeckis, 1984), Harrison Ford (Lucas, 1975),

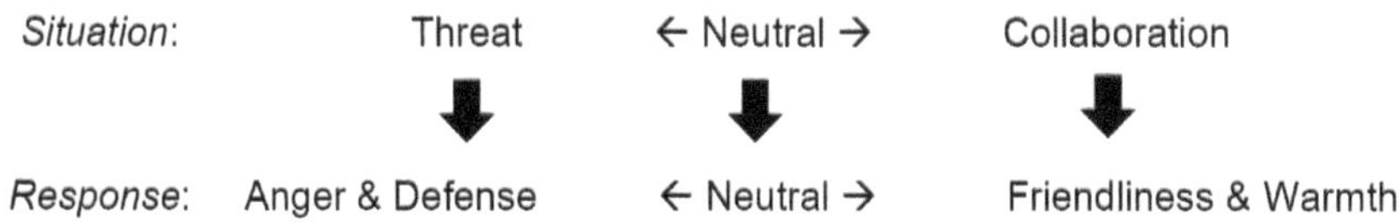

Figure 12.1 Territorial Logic of Anger.

and Meryl Streep (Pollack, 1985), seem to do a remarkably good job of displaying the appropriate *functional fluidity* of anger (Figure 12.1). When threatened, they show anger; otherwise, they are calm or friendly. Thus, some outstanding entertainment figures can function as readily available role models for displaying anger in an appropriate and controlled way.

Life can be thought of as offering a range of options, from threats on the left side of the range to neutrality in the middle of the range to opportunities for collaboration on the right side of the range. The basal, appropriate response is anger and defense in relation to threats, neutrality in relation to neutrality, and warm encouragement in relation to collaboration. That is an oversimplification, of course. But it does suggest a rational schema for experiencing and expressing anger.

From an ops perspective, people who function well with regard to anger are, first of all, respectful of themselves and have a lively belief in their rights. That provides an intellectual and emotional *warrant* for defending their territory. Second, they embrace the *reciprocity of that warrant*—they recognize the rights of other people to set up systems that meet other people's needs, to draw boundaries around those systems, and to have those boundaries respected (like an orchard or workspace). They are willing to negotiate all that, but in a spirit of empathy and fairness. That empathy and fairness means they rarely get into intractable or violently angry conflicts with others. Their self-respect and confident self-control, and their analytical, negotiative, and rhetorical skills means that others rarely take advantage of them.

When high-functioning people do feel their rights are being threatened, they can transition, if necessary, smoothly and convincingly to an angry position without losing composure or the ability to think clearly. As the other party backs down in the face of justifiable anger, a high-functioning person then can transition back to a neutral state with neither embarrassment nor regret, and maintain a strong and mutually respectful relationship.

As the noted lawyer Alan Dershowitz demonstrated (personal communication, May 2012), even in the midst of a conflict a clear display of anger may be ineffectual, even counterproductive, and shifting to a perfectly calm manner and negotiating logically may be more effective. Skilled negotiators know that the emotion is secondary and resolution of the conflict is primary (Fisher & Ury, 1981).

RECALIBRATION OF OPS COMPONENTS: ANGER MAKEOVER

- *Ideals*: It is not hard to idolize anger. Righteous anger is so impressive in an intimidating way, and, like punishment generally, it often works. There is, perhaps, nothing like a good explosion to get people's attention and assert at least the appearance of authority. But, also like punishment generally, if anger doesn't result in a fair arrangement, it typically doesn't *endure*, because people recover from actual aggression remarkably quickly, and then begin to plot their next move. As the strategist Edward Luttwak (1987) pointed out, aggression can be characterized by a somewhat self-defeating *action-reaction* pattern. It is something like bidding in poker. For example, Player A bids $1. Player B says, "I'll see your $1 and raise you $1." And so it goes, up and up. The general military pattern is that a dominant aggressive *measure* at Point A is eventually matched and trumped by an aggressive *countermeasure* by the opponent down the line at Point B. And so it goes, on and on (as the media frequently attest regarding political and international conflicts).

  A better approach on the whole might be that of Confucius, in praising the reliable middle way in life: "Who can practice tolerance and forbearance is not far from the way of the mean. He should do to others as he would have others do to him" (*Selected Collection*, 2006, p. 33). The ancient *Dhammapada* (1973, p. 68) suggested, "Overcome anger by peacefulness; overcome evil by good. Overcome the mean by generosity; and the man who lies by truth." The *Talmud* states, "A man must love his fellow man and show him proper respect" (Goldin, 1957, p. 66).

  Wisdom literature can provide a valuable counterbalance to the sometimes intoxicating, electrifying lure of rage and revenge. As compassion and generosity are genuinely respected and nurtured, anger can be experienced and communicated in a way that is rational and justifiable, and contributes to mutual respect and to personal and social harmony.
- *Cognition*: Clinically, unrealistic irritation with others often seems associated with unjustified claims of privilege and consequent expectations of exemptions from responsibilities. These presumptions are often derived from the zeitgeist—for example, parental attitudes, child-rearing guides, and even well-meaning educational and counseling systems. Unearned but claimed privileges have a powerful appeal, *a cognitive version of magic*. Like addictions, once established they are not easy to give up. And they are easy to defensively project onto other people, creating a maddening vicious circle, not too far from paranoia.

  Thus, a frequent clinical issue is that the belief in individual privilege and consequent exemption from normal requirements contributes to an eerie network of subtle and overt conflicts with the broader social system. People subject to early privilege-and-exemption conditioning tend to chronically experience *slights* both large and small from other people, because over and over the presumed and expected privileges and exemptions don't materialize. Life may be experienced as a chronic chafing process of insults large and small.

A way to deal with all that cognitively, clinically, is to start with a clean slate, to go back to a *beginner's mind* (Suzuki, 1970), to drop down to first principles and rethink the issue of talent, effort, rights, responsibilities, and rewards in life. A social premise, from the classical Greeks to Thomas Jefferson, is that all people are to be considered essentially equal, and at the same time all citizens are to be respected and admired. That beginning is very different from the presumption of differential privilege, an aristocratic notion, in America, abandoned at its founding (Herman, 2001). Starting with that equality premise, what might be called the *Horatio Alger principle* becomes operative—people can go as far as their talents and efforts will take them as long as they don't abuse other people's rights (Alger, 2008; Schaar, 1961). Those talents and efforts lead to a particular level of achievement. That level of achievement is associated with particular utilitarian benefits to society (Mill, 1863/1987), to organizations (Smart, 1999), to the larger good. The groups that benefit from the individual's efforts then reward the individual more or less in proportion to the individual's contribution to the larger good. The substitution of this *rational work-for-rewards* premise eliminates the alluring but ultimately frustrating *magical privilege* premise. That rational basis for operating thus eliminates the corrosive, illusory basis of irrational anger, which seems to be a burden not only to society but to many individuals struggling under narcissistic illusions.

- *Feelings*: It is surprising how often chronically angry people turn out to be ultimately grounded on a sense of being unloved, unappreciated, and disrespected (e.g., Beavers, 1977). Claimed special privileges then can be seen as *compensatory* (Adler, 1973). The attempt to claim special privileges is, ultimately, an attempt to feel better, an attempt to deal with feelings of loneliness and sadness, an attempt to acquire a sense of goodness, of belonging, of appreciation.

The ideal thing that happens in life, of course, is that a child is born into a family that is genuinely affectionate, admiring, caring, and encouraging—*healthy*. But it doesn't always work out that way (Bowen, 1978); people have children and get married for all kinds of reasons, some having to do with the proud nurturing of children, but some not. Excellent field research into *resilience* (Wolin & Wolin, 1993) indicated that some children find themselves born into families in which irrational disapproval, anger, and aggression are all too common. Some of those children, in wandering around the neighborhood and in attending school, find that good behavior aimed at the right recipients (e.g., friendly neighbors and caring teachers) results in exactly that affection and admiration that the children are seeking. Good feelings thus can be consciously generated.

- *Decisions:* The essential insight regarding decisions about anger is that, like many feelings, anger seems to thunderously rise up *sui generis*, like a glittering and well-armed giant abruptly looming overhead, in command of the field, sweeping all before it. But, naturally, anger doesn't possess a sovereign reality all its own. It is a complex of neurochemical events linked to other neurochemical events. Upon reflection, sometimes it makes sense, sometimes it doesn't. So, it is helpful to learn to think *about* anger, and to *make a decision* with regard to it, to, indeed, take command of *it*.
- *Action:* The essential action in anger is to assert rights in a thoughtful but brisk, justifiable way. That often is a four-step process of (1) review and assessment, then (2) a verbal statement of the issue and an offer to let the other person correct the situation voluntarily, (3) an attempt to negotiate the conflict, and then, if there is no reasonable negotiated settlement, (4) a clear expression of displeasure, and perhaps a warning of a plan of correction, should there be one, which is to say, some kind of punishment (e.g., legal action can be seen as a negative consequence).
- *Objectives:* Larger goals of living have been the subject of heated debate for thousands of years (Schaar, 1961). Is there really *anything* worth getting angry about? Some, like Stoics, would say not (Aurelius, 2006; Epictetus & Lebell, 1995). But then again, not many people have reached that degree of equanimity in the orchard of life. Most people are attached to their territories and the flow of supplies from those territories, and they don't brook intrusions cheerfully. So, the usual objectives in relation to anger are to defend territory while simultaneously maintaining good relationships with others. Good relationships, as the esteemed Professor Carol Gilligan (1982) proposed, are perhaps the primary criterion in living sensibly.
- *Communication:* Often there is an aura of paranoia and violence about anger, and that is not necessarily the kind of thing people want to advertise about themselves, or even know about. Therefore, it is very common for people to *repress* anger, as if it just were not there. As the psychoanalyst Karen Horney stated, "Repressing a hostility means 'pretending' that everything is all right" (1937, p. 64). But repressed, it can't be consciously dealt with, can't be refined and focused appropriately. Then anger can become a clinical difficulty. Or, if there is an angry outburst, that has dangers all its own.

Stanislavsky, the great Russian theoretician of the theater (Moore, 1965) argued that watching good acting, a good performance, is not completely unlike watching a person go through the real thing. Indeed, going through the real thing is not unlike acting. There is a surprising overlap. A performance is a performance, to some extent. Therefore, one way to frame anger is as a potential performance that can be *rehearsed.* A way for clients to rehearse clinically is to note an irritation, nothing overwhelming, and practice expressing appropriate irritation—sending the signal that a boundary has been crossed and that, as Queen Victoria supposedly put it, "We are not amused." In a process of desensitization and refinement, as clients learn to express and negotiate small irritations, larger ones can be attempted, until they can communicate anger appropriately over a wide range of seriousness and negotiate conflicts with clarity and impact, and a reasonable degree of comfort (e.g., in a marriage, or business), without the anger becoming overwhelming. *Rehearsals* provide the opportunity to polish feedforward so it is clear and convincing and to get more comfortable with feedback, which, in the heat of some moments, does not always conform to the ops template of *polite but definitive firmness.*

SUMMARY

Anger has its justification—defense of necessary territory. Dysfunctional anger may have its roots in narcissism and may eventually partake of paranoia. The ops approach is to respect personal territory as well as the territory of others and to express anger in a controlled, graduated way that is polite yet firm and clear. The goal is to resolve conflicts fairly and maintain positive relationships.

FURTHER CLINICAL CONSIDERATIONS

Surprisingly often, anger leads from and to other clinical difficulties, such as anxiety and depression. But to clarify the links, these issues need to be investigated individually.

Chapter 13

Anxiety

Journey from Fear to Courage

"I will show you fear in a handful of dust," said the poet (Eliot, 1930, p. 30). Perhaps *the* primordial fear is disintegration. And perhaps, therefore, *the* survival virtue is courage (Pury & Lopez, 2010). Fear and courage have a long pedigree in the human saga. Almost any great historical adventure that could be named likely revolved around *mastering fear with courage*—circumnavigating Africa the first time, traveling the Silk Road from China to the Middle East and back again, sailing to America the first time, discovering the South Pole, or landing on the moon. Those are the grand gestures. There is perhaps a parallel intellectual courage in asking about those huge and daunting realities of life; for example, the astronomer Carl Sagan's suggestion (1980, p. 4) that "Our future depends on how well we understand this Cosmos in which we float like a mote of dust in the morning sky" seems to breathe a rarefied air of immense, imaginative courage. Even ordinary challenges often call on courage—speaking up in class, asking someone out on a date, arriving at a new school, meeting someone important, applying for a job, getting married, buying a house, raising children, dealing with retirement, or facing death.

Initiation of action and *inhibition* of action are fundamental to the nervous system—the basic on-off arrangement, as indicated previously. *Fear* is a kind of inhibition of action, and *courage* is a kind of initiation of action. Thus, fear and courage can be seen as fundamental to the nervous system and fundamental to existence.

The technical term for fear is *anxiety*, of course, and it comes in all degrees, from mild uneasiness to panic. It is a fixture of formal diagnostic taxonomies. Anxiety is a core concept in psychodynamic thinking (Freud, 1930/1961); a multitude of what might be considered *surface-structure ills* actually may be due to *deep-structure anxiety*. Even some foundational behavioral thinking (Dollard & Miller, 1950) can be seen as dominated by considerations of anxiety, as in the conceptualization of neurosis as a fruitless alternation between approach and avoidance—danger is perceived in either direction, resulting in vacillation between options, an illusory safety rooted in nonchoice. Zen, too, deals with anxiety, as in Morita therapy (Reynolds, 1976). The discipline of calm breathing is about anxiety (Zi, 1994). Anxiety is one of the primary issues that brings people into psychotherapy.

DEFINITION OF ANXIETY

Anxiety could be defined within the ops system as *an uncomfortable, restless emotional state accompanied by a perception or even presumption of danger, a state that can escalate into panic, and terror.* Partly because the belief that there is a danger is often vague, partaking of a bit of paranoia, perhaps, like anger, addressing and resolving the threat in a forthright manner is frequently difficult. Sometimes a linguistic distinction is made between *anxiety*, as a pervasive but lower level of uncomfortable perception of danger, and *fear per se* that is sharper and at a higher level of discomfort and perception of danger. In the ops system fear and anxiety are used interchangeably, and contrasted with courage, their reciprocal.

THE FUNCTION OF ANXIETY: AN OPS PERSPECTIVE

Anxiety, like anger, is one of those states that at first blush seems to have no value. The kind of thing most people would prefer to kick to the curb. But fear is a universal human experience. It must have *some* value.

Perhaps a paradigmatic anxiety situation (a thought experiment) would be a young person wandering away from a village on a hot day carrying a big jar to get some water for the village from a stream. The winding trail down the hill to the stream is long and dusty and rocky. The young person wearily collapses at the bottom of the hill under a shade tree for some rest. At that moment warriors from a rival village appear on the hill across the stream. Suddenly the young person is terrified and thus *energized*, and takes off straight up the hill at top speed while loudly yelling a warning.

So, there *is* a certain value to fear. It focuses the mind wonderfully, generates instant priorities, selects objectives, identifies paths, and gets people moving *right now*. Thus, it has survival value. As with so many emotions, *managing it well* is the challenge.

TYPICAL CLINICAL COMPLAINTS REGARDING ANXIETY

Though anxiety initially has a motivating effect, spurring people to action, *chronic* anxiety, detached from conditions, is different. It is debilitating. Plus it is uncomfortable—a nervous, at-loose-ends feeling at best. There is the physical tension, which is tiring, and the consequent restriction of breathing and low oxygen, which contributes to low energy and a sense of panicky, suffocating discomfort. But then there is also the pervasive sense of being surrounded by dangers that cannot ever quite be seen or touched, and which therefore are difficult to handle. It is frustrating and disorienting. There is what might be called the *Neverland illusion*—a perpetual, intuitive, and powerful sense that the best way to deal with threat is *avoidance or escape*. Naturally, in dangerous situations, fleeing *is* often the best option—it is relatively quick, and many times complete; safety is achieved. But, once avoidance and escape occur and are successful, they are reinforced, and evasion can evolve into a pervasive default strategy. People can become lost in that isolated space, that island of imagined security nonetheless populated by hidden malevolence difficult to defeat, like Capt. Hook in Neverland.

From a larger perspective it turns out, as many people know, most things in life, especially in urban life, which are initially *alarming* eventually turn out to not be actually *dangerous*. The issue then is not dealing with the *danger*, which appears to be there but often is not actually there; rather the issue is dealing with the *alarm* itself (Ellis & Harper, 1975; Epictetus & Lebell, 1995). It turns out that *habituation* occurs in response to scary things *if those scary things are routinely faced* (desensitization). For example, standing on heights, going up elevators, meeting new people, or giving speeches become more comfortable if they are approached routinely.

And that is the clinical conflict. If the client would face stressors in a graduated, controlled way, the client probably would habituate to the stressor, and the alarm reaction slowly would fall to a reasonable and manageable baseline. But because the client has routinely avoided stressors, instead of being *de*sensitized to the stressors, the client has been *sensitized* to the stressors. Thus, what start out as minor fears can evolve into massive terrors (e.g., agoraphobia), resulting in stressors being even harder to face than they were initially.

Overall, the ops approach to such a *symptom picture* is, as usual, largely indirect—a systems approach. The ops model assumes that the experience of feared stimuli and the coping with feared stimuli are guided by the personality as a system. Therefore, the ops way to help reduce anxiety is to strengthen the personality in the area of courage and forthrightness, component by component. A beginning to that process is to consider what a markedly courageous person would look like from an ops perspective.

PORTRAIT OF HIGH-OPS PERSONALITY IN RELATION TO ANXIETY

A high-functioning ops person respects fear. Fear is a signal of a danger in the environment. Fear targets action and energizes the entire system. Fear is a useful motivator. Thus, fear is the personality's *friend.* Sir Chris Bonington (1981), the professional mountain climber and adventurer, suggested that all serious adventurers feel fear and respect their fear. They want to keep an eye on it, as it keeps them informed about danger. It is because of fear that they so meticulously prepare, it is because of fear that they keep their minds so alert and rational despite pressure, and it is because of fear that adventuring is so much fun. This perspective—that fear is something *useful* and even *interesting*—amounts to a *therapeutic reframe.*

High-ops people maintain a close, interactive correlation between the signal, anxiety, and the danger to which the signal refers. As anxiety rises, the correlate is checked. If indeed there is a danger, it may make sense to maintain a high level of motivating anxiety to help deal with the danger. If anxiety is aroused, but a check indicates the actual danger is minimal, then there is a conscious attempt to reduce the anxiety with reassuring self-talk, such as "It's no big deal; I can cope." High-ops people work *with* their anxiety as they navigate the risks of life, and maintain a positive attitude toward risk—not only is risk often involved in great enterprises, it is part of the excitement of living, something like a sailboat heeling over as it skims briskly over the water in a tight race (e.g., the pediatrician Stuart Walker on the thrill of Olympic-level competition, 1986).

RECALIBRATION OF OPS COMPONENTS: ANXIETY MAKEOVER

- *Ideals:* It is clear from that cornerstone of Western culture, the *Iliad* (Homer, 2004), that courage was an explicit part of the classical Greek world. "I speak of champions among men of earth, who fought with champions, with wild things of the mountains, great centaurs whom they . . . overpowered" (p. 14), and "Athena . . . that dazzling goddess went to stir . . . each man in his heart grew strong to fight and never quit" (p. 44). These statements suggest that courage was consciously embedded in the mentality of the populace. As courage rises, naturally, fear diminishes, since these are reciprocal states of mind. The Asian *Book of Five Rings* (Musashi, 1974) famously canonizes instant courage in the face of danger.

  In similar fashion, individuals, families, and groups can consciously adopt ideals and values that prize courage. So, that could be the first corrective step—making courage and courageous role models a part of the personal pantheon. The next steps could involve watching movies or shows, or reading books that canonize courage. Interestingly, and perhaps instructively, spiritual traditions and materials often showcase courage—from the Hindu *Rig Veda* (O'Flaherty, 1981), to the Stoic *Meditations* (Aurelius, 2006), to the Shaolin Temple's spectacular kung fu practitioners (Polly, 2007), to Michelangelo's resolute *David,* to Athena, goddess of wisdom as well as war (Bowra, 1957). Those traditions suggest that development of qualities such as serenity, profound compassion, and liberation from petty concerns calls for a commitment to high ideals not unlike combat courage.

- *Cognition:* A great deal of mental life can be seen as being comprised of three cognitive processes: (1) recognizing the systems that generate relevant events, (2) keeping track of the systems and the events, and (3) making predictions as to what events likely will be generated by any given system at any given state in the future. That amounts to what might be thought of as "*systemic-probabilistic reality.*"

  Surprisingly often, clients who feel buffeted by anxiety have a disconcertingly vague sense of how things work, what is likely to happen, and how the clients conceivably might steer events in a preferred direction. Thus, they frequently find themselves haunted by ghostly and unsettling premonitions. By contrast, people who rarely feel buffeted by anxiety often have a precise, tangible sense of what is going on and what will happen, and how to make things happen—*they have a good grasp of systemic-probabilistic reality.*

  Therefore, one approach to reducing dysfunctional anxiety is to work on clarifying thinking in an obvious way, getting in the habit of asking such questions as, "How does this system work?" "What does it produce?" "Can the system be adjusted so that it produces different things?" "What can I do to make those adjustments?" Just learning to think in this fundamentally *operational* fashion increases a sense of control, builds confidence, and reduces dysfunctional anxiety.

  For example, when people start a job in an organization, they often are mystified by who gets promoted, and how. That perception engenders a certain anxious helplessness. But as employees gain information over the course of a year or two, typically it becomes clear that although politics and executive suite personal preferences are *elements* in the promotion process, the criteria are largely a sensible composite of attendance, attitude, competence, camaraderie, and confidence, as well as executive potential (Toropov, 1999). As employees get a better grasp of the way the system *actually* generates promotions, they not only gain confidence in the process, they gain some control over their own future by becoming the kind of person who would be a good candidate for promotion.

Another cognitive approach, as alluded to previously, is the *reframe* in which dealing with the issue is conceptualized as a *game*, an *adventure*. That can work clinically if at every step the threat level is kept low, and comfortable mastery is the goal. The process of learning to deal with a feared stimulus then can be seen as something like learning to sail a boat—a challenging situation that, while disquieting at first, is practiced until it becomes fun (e.g., Walker, 1986).

- *Feelings:* If that works so well, it might be fairly asked—why doesn't everyone do it? The answer, of course, is anxiety itself. Understanding situations as they really are and seeing where they likely are headed can be scary. For example, facing family secrets or repressed thoughts; dealing with sometimes dicey prospects in school, dating, or work assignments; candidly considering the state of a marriage or of health, or the cost of ambition can be unsettling. For people of shy and sensitive temperament, the process of *precising—clarifying things right down to their exact elements and actions*—can be alarming, even overwhelming. Facing such stressful situations slowly and relatively comfortably is a core clinical process.
- *Decisions:* One of the verities of living might be called the *coping paradox*—things need to be faced, but then again, it is not that good of an idea to just plunge in willy-nilly and slosh about. So, the basic decision with regard to anxiety is to *commit* to facing things, but in a *graduated*, step-by-step fashion so that desensitization can occur. That thought alone, that exposure will occur but it will be controlled, is often reassuring on both counts—things are going to be dealt with, not just left to run amok, but they will be dealt with in a relatively comfortable way.
- *Actions:* Once the basic decision is made to face a feared stimulus, an *exposure plan* can be developed. The stimulus can be analyzed, and threat can be graded, so the client is aware of the least threatening part, type, or phase. Then the exposure can be organized into desensitization steps, so the least threatening or most interesting part, type, or phase can be dealt with first. The reframe attitude brought to the action is one of curiosity, intrigue; how to cope with the feared stimulus is initially something of a mystery, and mysteries are interesting. Modal self-statements might be, "How will I feel as I am approaching this situation?" "What will I think?" "How will I handle my feelings and thoughts?" "What looks like a good first step?"

For example, meeting new people can be stressful. Anxious thinking is often radically dual: black or white, fight or flight. For example, youthful, shy people coming to a faraway college might think in terms of either being lonely *or* going to a school activity and cheerfully partying until all hours in a huge crowd of strangers, but neither of these black-and-white alternatives is comfortable. But almost all options can be structured as a graduated, stairstep range within anchoring extremes. On the left would be the loneliness option. On the right would be the uninhibited socializing option. But in the mid-range are graduated increases in stress and reward that are manageable.

A good beginning for a lonely, shy student might be to go to the school cafeteria every morning early and become friends with the cafeteria staff, who are required by their jobs to be reasonably friendly. That then forms a cushion of confidence. Then, it is likely that other people show up regularly in the morning to eat and study, and those people become mutually familiar. A glance or a smile is fairly easy to produce, even in anxiety. It is not very hard to sit near someone who looks friendly. Quick questions or comments, such as "Are you taking introduction to history?" or "Aren't the omelets good here?" are not too risky, and will usually elicit at least a polite response. Professors and department staff can be quite friendly and helpful, especially to new students, and can comprise another support group. Dormitories and student apartments, not to speak of classes, sports teams, and student organizations, also offer companionship that can be established safely and over time, bit by bit, starting easy and moving up the intimacy and hedonism scale, until social relations are fun, interesting, and minimally stressful.

The same is true for other situations, such as moving into a new neighborhood, starting a new school, starting a new job, joining a new organization or club, embarking on a difficult new project or acquiring an intimidating skill, such as speaking a foreign language, debating, or playing a musical instrument. The key to desensitization and skill acquisition is to conceptualize the challenge as a stairstep of difficulty, and commit to starting at the bottom step, the easy step, and working slowly upward until competence and comfort are achieved.

A step up from all that is to consciously, physically *embrace the adventurous life*. That can entail joining a sailing or hiking club, learning to kayak, traveling around the country in a car or to foreign countries. The possibilities of fascinating adventures in life is really limitless. Each conquest builds confidence.

- *Objectives:* As with most projects, specifying the objective makes accomplishment of the objective much easier. But anxious people are often reluctant to specify objectives. There are a number of reasons for that.

  As suggested previously, if the objective is not specified, it can often be avoided. Anxiety-provoking challenges are often avoided because of an underlying fear that personal emotional resources are on the verge of being tapped out, and a vague sense that one more failure may plunge the whole system down into the emergency red zone of functional insufficiency—catastrophic failure. But if the objective is never specified then there will be no failure to reach it, and the system is safe. Again, that safety is a little illusory.

The second reason, that very logical one, is that specification of what *will* be achieved simultaneously specifies all the other things that *will not* be achieved. Selecting A usually means that B, C, and D will not be selected. Contemplating all those potential losses is anxiety-provoking! It is often a little hard for highly sensitive people to fully grasp that if *no* objective is ever selected, then it will not be just *most* of the options that will not be achieved, but *all* of them.

The third and by now somewhat familiar reason that anxious people are reluctant to select objectives is that clarifying *ends* has a tendency to simultaneously clarify *means*, and that can be anxiety-provoking. As long as the ends remain vague, the means remain vague. If an anxiety-prone person were talking, say, to the writer-adventurer Ernest Hemingway at a safari camp, and Hemingway said, "Let's take a walk," that would be, more or less, fine. But if Hemingway said, "Let's take a walk *to the top of Mount Kilimanjaro*," that might be, more or less, alarming. Anxious people often experience ends and means specifications that way—suddenly, a huge demand, complex and difficult, that could easily spiral down into failure and ignominy. Again, in anxious people there is a sense of personal resources almost being tapped out and a sense of everyday existence being conducted on a razor-thin margin of safety.

A first way for clients to approach all this is for them to conceptualize objectives in good ops fashion *heuristically*, in an as-if, experimental mood. An encouraging example of self-talk might be, "Well, maybe I'll just try this particular objective here and see what happens." That creative, experimental attitude reduces that claustrophobic *uh-oh* fear of plunging into an irrevocable choice with no way back out, and permanent loss of all other options (i.e., *piloting* an objective).

The second way for anxiety-prone people to approach objective-setting, similar to what can be done with action steps, is to conceptualize objectives, not as total black-and-white options, but as graduated stairsteps—easier objectives first, and then, as success and confidence are built, moving on to progressively more difficult objectives one by one. A reversal of the usual Olympic medal progression may help to conceptualize the steps upward—clients can first attempt easier bronze-level objectives, then the silver-level objectives, and finally the tough gold-level objectives, if those are desired.

- *Communication:* In a style parallel to safety-oriented thinking patterns, anxiety-prone people often communicate to themselves and other people in a manner intended to protect themselves from failure and loss, while simultaneously intended to convince themselves and others they are decisive and bold. But, usually, the style of communication just compounds the anxiety.

The communication pattern, at a deep-structure level, is vague, because the thinking is vague, because vagueness is experienced as protective in a hostile environment. It is not quite clear what they want or do not want, are planning to do or are planning to not do, doing or not doing. But they are frequently subject to negative feedback about all that. Therefore, at a surface-structure level they often adopt a tone and language that convey an impression of bold decisiveness. Their real thinking and feeling thus becomes a shadow of their public persona. This compensatory style is very confusing, and makes progress in reducing anxiety and advancing toward true goals difficult.

Gestalt therapists (e.g., Perls, Hefferline, & Goodman, 1951) and some short-term psychodynamic therapists (e.g., Davanloo, 1992) recommend *leveling, engaging in direct candor* so that everyone knows what is going on, who is who, and what they plan to do in reality. It is not necessary to be like that all the time, of course. Tact, negotiation, and role-playing are essential parts of a successful social system (e.g., Goffman, 1974; Nierenberg, 1968; Post, 2014; Satow, 1921/2006). But coming to terms with true thoughts and feelings, and being able to communicate that inner reality in a reasonably valid way is very helpful in overcoming chronic anxiety.

Being able to communicate in a direct way takes practice, and needs to be practiced in a supportive, understanding, encouraging environment (such as psychotherapy). And, like action and objective-setting, it can be done in a stairstep fashion—some honesty at first (bronze level, so to speak), then direct leveling (silver level), and finally tactful candor (gold level). Simple difficulties and limits can be discussed first, more challenging ones later. The eventual goal is to be able to have frank discussions at will, such as in a relationship, or when parenting.

SUMMARY

Anxiety can seem like a pervasive and mysterious disorder, of ghostly origins and opaque destiny. But the ops interpretation is that it follows a particular path of development relying excessively on avoidance of stress. The ops approach to build up the strength of the basic personality components, aiming overall toward a more courageous, adventurous life within a game-like spirit.

FURTHER CLINICAL CONSIDERATIONS

An anxious temperament does have its advantages; people who are outstandingly sensitive and caring, artistic and intelligent are often anxious.

Nonetheless, chronic anxiety can limit options in life. Being highly sensitive, while perhaps contributing to creativity, sophistication, and emotional depth can, if taken to an extreme, negatively impact self-image and social relationships. And that can veer into depression. Therefore, understanding and reducing chronic anxiety usually also means understanding and reducing depression.

Chapter 14

Depression

Journey from Darkness to Light

Leo Tolstoy, in his immense and timeless novel *War and Peace* (1993), presented the protagonist, Pierre, as depressed (or at least he seemed depressed). He was a member of the Russian nobility and appeared to have just about everything going for him, as Tolstoy describes him. Pierre particularly enjoyed visiting the home of the Rostovs, portrayed as an aristocratic though financially compromised family. He enjoyed those visits partly because of one of the Rostov's daughters, Natasha. She was described as a charming young woman with a bright mind, a fluid wit, and a rollicking warmth, who displayed genuine, cheery affection toward everyone, high or low, young or old, rich or poor. In the 1950s version of the movie (de Laurentiis & Vidor, 1956) Natasha was played, not surprisingly, by Audrey Hepburn. Pierre absolutely bathed in the light of the human spirit that Natasha radiated. Nonetheless, that experience, that relationship somehow could not quite illuminate his overly thoughtful, brooding existence.

Pierre seemed caught, or allowed himself to be caught, in a strange and subtle web of artificial class consciousness, exalted but remote expectations, and actions detached from personal meaning, all within a chronic sense of fear, of defeat, of not belonging, and of being lost. Pierre hungered for the kind of bold, imaginative, revolutionary brilliance symbolized by the figure of Napoleon. And yet Pierre experienced great difficulty in making the *critical shift from wishing to doing*. He experienced great difficulty in creating a genuine connection between his inner being and the outer life he lived. It is not surprising he appeared depressed, because he moved, something like an actor in the wrong play, through a world that did not adequately match his real personality or his real aspirations, a world that asked from him things he was not interested in delivering, a world that rewarded him for things he did not care about and punished him for things he deeply believed in.

War and Peace (1993) is an almost overwhelming book about the fruitlessness of war. The literary massiveness can be seen as symbolic, subliminally, of the heroic struggle of a depressed person against an inner reality that does not quite fit and against an outer reality that does not quite fit. The depressed mood of an individual is an index of the mis-calibration between, as it were, the inner rock and the outer hard place, between the life wished for and the life lived. Lifting mood, strengthening personality, *redesigning and reconstructing the systems* from which the moods arise, that is, *reorganizing personal operations*, is difficult, and at times seemingly impossible. This is akin to systems engineering, an enterprise that is holistic as well as meticulous (Roadstrum, 1978).

In a parallel example, *Mission Impossible* (Geller, 1966) was at one time a popular television show. It always began with the protagonist, a suave agent, listening to a secret tape recording in which some impossible-sounding mission was proposed, but which also offered the thoughtful caveat, "Should you choose to accept it . . . ." That is about the position depressed people are in phenomenologically. The challenge of change can seem massive and impossible, yet they choose over and over to try to deal with it.

DEFINITION OF DEPRESSION

Depression could be defined within the ops system as *low energy resulting from discouragement* (sometimes leading to a vegetative state). The term *discouragement* indicates there is a cognitive component to depression—a belief that effort will not lead to reward, an intuition of being trapped. The root of the term—dis-*courage*-ment—also suggests there is an anxiety component to depression, a reduction in courage. This is something like Martin Seligman's concept of *learned helplessness*—a sad fearfulness that follows repeated punishment with no apparent way to escape or to take control of the punishment (Wilmshurst, 2005), and is something like the posttraumatic stress disorder (PTSD) experienced by children who come to feel a "Dread of being flooded" with feelings of trauma (Dwivedi, 2000, p. 3).

It also bears a certain resemblance to the idea of *learning-based representational catastrophe*, in which a part of the brain that ought to be maintaining proper mental distinctions and therefore maintaining control of option selection has become *fused* and therefore has become ineffectual in managing a critical sector of reality (Schwartz & Begley, 2002). That does seem to happen when people repeatedly strive for goals and fail, such that the means, ends, and creative alternatives (process and hope) all get fused into one overriding sense of ineffectual pointlessness. The corrective idea of *neuroplasticity* is that people can decide to make a difference in the way the brain works, and can make small changes repeatedly, such that new pathways are constructed and thus new opportunities created (Nelson, de Haan, & Thomas, 2006). That perspective provides *hope*.

THE FUNCTION OF DEPRESSION: AN OPS PERSPECTIVE

Depression is another one of those states that most people would probably like to kick to the curb and never see rise up again. However, there *is* a point of view from which depression makes sense.

Mania—energy, optimism, and good feeling carried to a self-defeating and sometimes psychotic extreme—is usually contrasted with depression, as in bipolar disorder. But it may be more useful and ultimately more accurate to contrast *enthusiasm* with depression.

People have needs as well as desires and the bolder and more far-flung personal dreams. People usually are thrilled about meeting important needs and fulfilling major dreams. The *thrill* then is a feeling of high excitement and deep satisfaction. Multiple thrills create a life of enthusiasm. Those positive feelings are reinforcing—the association is strengthened between doing good things and feeling good. So, all that results in living what might be called an electrified, elevated *Zarathustra existence* (drawn from Nietzsche, 1966); living in a *flow* of comfortable, secure focus (Csikszentmihalyi, 1990); that is, functioning at a *high-ops level*, where life is largely under control, life is largely successful, and, though not to the point of being manic, feelings are largely positive (e.g., Fredrickson, 2009, and the field of positive psychology).

Behavior control systems seem to work best (regrettably, perhaps) when behavior is associated with both positive stimuli (the famous *carrot*) and, at least to some degree, negative stimuli (the famous *stick*). For example, in organizations, positive stimuli, such as bonuses, perks, and promotions do help a great deal to motivate people and keep them on track. But, realistically, it probably also helps that everyone simultaneously knows that negative performance reviews, dismissal, and negative references can be associated with sub-threshold performance. In police work this is known as the *nice-cop/tough-cop* approach. In international relations this is *diplomacy* backed by *force* (Satow, 1921/2006).

The mind can be seen as working similarly. The human system wants to survive. That means doing good things—being successful, and not just once, but over and over. To repeat the success the successes need to be rewarded. One of the best rewards and, in a way, *the* fundamental reward is *feeling good*. (Perhaps the keystone book in CBT is titled *Feeling Good*; Burns, 1999). So, that is the positive stimulus within the system. But the system is *aware* that positive feelings *alone* might not be enough to survive. That might be why Pollyanna, that wonderful and brave fictional personification of an *exclusively* cheerful outlook, has disappeared from the national scene (Porter, 1913/2003). Therefore, the mental system does and perhaps must include *negative feelings*. Negative feelings, like sadness and loss, feel unpleasant, they *hurt*. Thus, they function as *punishers*. And thus they also function as *motivators*. They help insure that effort will be directed persistently and imaginatively, despite discomfort, toward success, and missteps will be scrupulously avoided in order to avoid painful *failure feelings*.

This cost-benefit, pro-con, win-lose, happy-sad combination runs throughout life and seems to be one of those realities of the human system—*the thrill of victory and the agony of defeat*, as the saying goes; part of human greatness. That is to say, if it did not *matter* so much, would anyone try so hard? That *mattering* means that success feels great, and not succeeding is painful.

There is an old joke about a plumber, a joke that carries a lesson about *prominent effects being supported by hidden systems*. The plumbing in a commercial building was not working. Several plumbers had been called in, and none could fix the problem. Finally an extremely expensive specialist plumber was called in. The charge to fix the plumbing problem was $1,000. The owner of the building and the high-priced plumber, who was holding a wrench, wandered around in the basement of the building, looking up at the silent plumbing system. Finally, the plumber peered at one particular valve, reached up, and tapped it with the wrench. The plumbing system kicked in, and water could be heard running through the system again. The owner exclaimed, "That little tap cost me $1,000?" "No," said the plumber. "The *tap* was only $1. *Knowing where* to tap was $999."

Systems are a little like that, including systems of knowledge. There is some kind of an output that by itself seems isolated, almost *sui generis*. It could be called the *iceberg paradox of the mind*—the impression of a small, jewel-like attitudinal and behavioral formation floating all by itself across the surface of reality, whereas, in fact, the attitudes and behavior have systemic sources—the other nine-tenths of the iceberg, within an individual, the personality.

Structural modeling (e.g., Cozby, 1997) captures this idea in diagraming as to how a set of input variables correlate with, and appear to contribute to, one or more output variables. *System analysis and design* (Yourdon & Constantine, 1979), as well as *systems thinking in relation to families* (Watzlawick, Bavelas, & Jackson, 1967), are all about the way that complex aspects of a situation feed into results. The results, then, are a *function* of the system; to understand the results it is necessary to think about the system that contributes to the results. In family therapy the manifestation of difficulty (the tip of the iceberg) is often termed the *identified patient*, the *IP* (Satir, 1967).

In regard to depression, the depressed feelings could be considered the IP, and the rest of the personality the contributing system.

Rain, as a natural system example, is the result of numerous elements, such as temperature and clouds, combining in a particular way. The appearance of sunshine, similarly, is the result of elements combining in a particular way, such as the temperature rising and the clouds dispersing. Weather systems are often compared to moods—dark or sunshiny.

To understand human moods, it is necessary to take into account the personality as a system, a system that generates moods. To try to improve mood, it seems reasonable to try to strengthen the system as a whole.

TYPICAL CLINICAL COMPLAINTS REGARDING DEPRESSION

Clients usually seem to present regarding depression with concerns in two planes. In the conscious plane, there are painful experiences of depression—low energy, low self-esteem, helplessness and hopelessness, vegetative tendencies, and sometimes fantasies of self-destruction as a way out of what are often viewed as a series of traps, an unavoidable succession of disappointments. In the subliminal plane, or even further back, in the unconscious plane, there often lies a powerful and uncanny fear of shaking off limitations, an intense anxiety about seizing opportunities, what might be termed a *carpe diem phobia—a fear of losing the illusion of having magical control over life*. Ironically, the conscious concerns about and experiences of depression may serve to buffer the personality from the *essential operational angst—a recognition that needs and desires call for sustained, focused effort*. That tends to be the depressed *perception* of operational reality; from a depressed point of view, it seems like life is unmanageably demanding and difficult, and ultimately unworkable and unsatisfying.

Possibly, something happened in the early years to engender this proclivity in which people dream dreams, but then do not take the probable paths to those dreams, and get stranded in pure wishes. A possible reason is that "longing for . . . magical power" noted by the psychoanalyst Alfred Adler (1973, p. 7), that self-defeating struggle to cope with both feelings of inferiority and compensatory cravings for superiority. A second possibility is the case in which children are secretly trying to please, and not threaten, a loved but insecure parent leading to a *fear-of-success* state of mind. Thus, failure in an obvious, public frame can disguise a striving for success in a larger, hidden frame—the frame of the frame (Goffman, 1974). In either case—a craving for magical power or a fear of success—that struggling, self-defeating, failure lifestyle hurts. It is depressing.

The preliminary ops approach to all this begins with three proposals for clients. First, to consider as a heuristic premise *hope*—a firm if tentative belief that *things can get better*. Second, to consider that the primary place to focus attention is not so much on the experience of depression per se as on the personality as a whole, on the content and functioning of the basic seven personality components. As those components are *recalibrated toward courage and enthusiasm* the depressed feelings are likely to be reduced and to become more manageable. Third, to realize that *none of that has to happen all at once*. A careful, comfortable contemplation of the overall situation is helpful in easing toward the goal of confidently dealing with the self, with life, and with depression. Then the components can be considered one by one, and plans can be developed for inclining the components and the personality as a whole *away from fear and sadness toward courage and enthusiasm—joie de vivre*. Reviewing a portrait of a person who, while sensitive and thoughtful, is not bothered excessively by depression, may be an inspiring first step toward moving away from depression.

PORTRAIT OF HIGH-FUNCTIONING OPS PERSONALITY IN RELATION TO DEPRESSION

A high-ops person is sensitive to the needs of others and assumes some kind of basic *rational contract in living*. High-ops people assume that friendliness linked to generosity is the *standard* by which living is evaluated, like the inspiring Greek literary character Menelaus when he was stranded and hungry in Egypt: "I refuse to be frightened off . . . no man alive could be so uncivilized as to deny me food" (Euripides, 1981, p. 38). People who are resistant to depression have *faith* in themselves, in other people, and in life. They are nonetheless *realistic* about life. They are rarely seriously surprised by outcomes. In considering the basic fairness contract of life they understand that things they want are premised on things they are willing to do. Their effort is smoothly, intuitively calibrated to their goals.

As realistic as they are (they are a little hard-nosed), they tend to see people and life in a positive light (they are forgiving). They tend to give other people, and life, the benefit of the doubt and to generously overlook flaws and foibles, and take in stride little disappointments.

Depression-resistant people focus on strengths, not weaknesses. They tend to be hopeful and optimistic as opposed to doubtful and pessimistic. They focus on what works, even as they take shrewd account of what doesn't work. When things don't work out as expected, they are not emotionally overwhelmed; rather, they regroup, reanalyze the situation, and plan a new approach likelier to work. They reengage with the situation with confidence and flexible creativity. They push through to success, rather than becoming mired down in self-doubt and self-criticisms. If a need or desire absolutely begins to look like it can't be achieved, they withdraw and try something different altogether, with little loss of self-esteem or momentum. Their *joie de vivre* goes deep, so that even as they sensitively register the tragedies, heartbreaks, and injustices of life, they personally remain stable, cheerful, and proactively engaged. They help where they can, but continue with their lives. Grief occurs, but they bounce back; they are resilient, partly by conscious intent.

RECALIBRATION OF OPS COMPONENTS: DEPRESSION MAKEOVER

Developing a positive outlook on life, along with energy and creativity, is not unlike growing a garden. A garden starts out a piece of property. Seeds or small plants are put in the ground. They are fertilized, watered, and pruned. And (if it isn't the dead of winter in New England or the height of summer in Las Vegas) they live. They bud, they blossom, they thrive. It all flows from the gardener's *desire, intent, and effort*. The effort doesn't have to be massive, unlike *War and Peace* (Tolstoy, 1993), it just has to be *consistent*.

It is the same with nurturing the components of personality. It can be done in easy steps, without pressure or undue concern about possible failure. Personal development, as William James (1890/1950) and Stephen Covey (2004) pointed out, is largely a matter of developing the *habit* of maintaining helpful thoughts and carrying out helpful activities. Developing new habits is possible because of the plasticity of the brain and its ability to lay down new pathways based on new wishes and intentions. Then, once a process has been developed into a habit, it is much easier to pursue consistently. Indeed, it becomes hard *not* to do it that way. It becomes a part of the personality.

- *Ideals:* The famous, heart-rending, and much-beloved Puccini opera *La Boheme* (Auspitz & Dornhelm, 2008) is about young artists in nineteenth-century Paris who are so desperately poor that they barely have enough to eat, have almost nothing with which to heat their frigid atelier, and finally lose one of their group to a fatal illness. However, the story is *so charismatically romantic* that at the conclusion it is hard to not crave a voyage to Paris to become a young, poor, starving bohemian artist. Just so, it is not very difficult, especially for people of a sensitive nature, to slide into identifying with and then idealizing sadness and discomfort—it helps to feel normal and take off some of the pressure to cope. But, other than a trip to the opera, it is probably more desirable in the long run to consciously adopt ideals of courage, confidence, energy, persistence, and cheerfulness and to derive from those day-to-day values of resilience and responsibility, curiosity, and satisfaction. Numerous sources, such as songs, movies, and books, can supply inspiring myths, stories, and role models. For example, to develop intelligent, cheerful resilience, an upbeat biography of Florence Nightingale or Alexander the Great probably would be a better bet than something magnificent but disturbing by Tolstoy or Puccini.

- *Cognition:* Ultimately, what is depressing about depression is that perceived existence is *lower* than desired reality. That sense of being at the bottom of a well is vastly compounded by a sense that rising out of the dark and into the light is hopeless, impossible. Here again, the emotions can be seen to be a result of a particular *formulation* of life, based on a *presupposition* that life is *supposed* to be different than it is. That creates a cognitive illusion of being *down*. As the presupposition is addressed, as it is clarified that the client has *built up* the ground of expectations above the starting point, *creating* the psychological well, the client can begin to dismantle the presupposition, dismantle the perception of being at the bottom of a well rather than merely at a starting point, bringing the analogical top of the well *down* to the actual starting point, then can begin to perceive reality as it is, and can begin to address desires and objectives in a more forthright manner. Dropping *depressogenic presuppositions* is greatly assisted by consciously making a habit of being factual, logical, and probabilistic, and putting a positive spin on all that—another habit.
- *Feelings:* Depressive feelings can be powerful. Bipolar disorder, for example, can be frustratingly difficult to treat even with carefully designed regimes of medications and therapeutic contexts. In more tractable cases, less clearly biologically governed and more apparently functional and psychosocial, the key rule might be something like one of the major guidelines in dealing with an allied disorder, panic—*don't get panicked by panic*. The idea is for clients to not get *entrained* by the negative habits of mind that typically characterize depression. Rather, they should *stop* and deal with the initial stimulus in as forthright a manner as is manageable. In relation to depression the therapeutic slogan would be—*don't get depressed by depression*. In other words, the guideline would be for clients to try to avoid compounding the initial disappointment by getting immersed and then lost in that vague, chilling, and often dramatic horror that is the emotion itself.
- *Decisions:* That *stop-and-consider process* is a decision that can become habitual if a conscious attempt is made to exercise it. That decision *interrupts* descent down the mood curve from initial recognition to the throes of sad mood, which, once arrived at, can be difficult to control.
- *Actions:* In a way, the key in depression management is *taking action*. That, of course, is the ultimate conundrum for a vegetative state that, by

definition, is weighed-down nonaction. The trick is to think in terms of *antidepressant routines—short sequences of actions aimed at very reachable objectives that always begin with easy and appealing first steps.* Every achievement (such as meeting needs) can be seen as standing at the top of a stairstep of sub-achievements, of prior, supporting objectives. Achievements can always be broken down into easier routines, can be started at an easier step. Thus, a successful life can be seen as being based on relevant, manageable routines, an elevated stage arrived at through easy steps forward and upward.

- *Objectives:* In relation to depression, which can be quite debilitating, objectives need to be kept manageable—specific, but quite feasible. That means learning to ignore the compensatory siren call of grand and mesmerizing plans. That, as with Homer's Odysseus (1963) sailing past the island of the sirens, can call for some discipline and determination. It helps to have the insight that, in the long run, it is a far more satisfying (less depressing) arrangement to set meaningful but modest goals, and then more or less comfortably easing toward the habit of routinely reaching them. This is a measured and slow approach, but likelier to lead to success, keep at bay anxiety, and improve mood.
- *Communication:* Language appears to have a surprisingly powerful effect on thought and mood, an effect that is often not conscious. This is the famous Whorfian hypothesis, discussed by the psychoanalyst Jacque Lacan, who reiterated that words are the primary medium translating observations into thought and that *letters* formulate reactions to thought (Ragland-Sullivan, 1986). Therefore, it seems therapeutic regarding depression for the client to consider adopting the following tripartite communication scheme. (1) Choose language that has a positive, hopeful theme, so statements sound upbeat and cheerful, such as: "This looks good," or "This might work," or "Things are coming along." (2) State issues less in terms of emotions and more in terms of actions and objectives. (3) When talking to others about life, be careful of promising or expecting too much; in speech stay limited, modest, probable, and doable. This helps relieve some of the burden of unrealistic expectations, which frequently are a source of depression (Ellis & Harper, 1975). This results in language that is *encouraging, goal oriented, and feasible.*

SUMMARY

Depression can be very difficult to deal with. But the ops approach is to look at those aspects of the mind that are depressogenic, and to develop and adjust them (recalibrate them) so that they are overall more uplifting. The key process is to reduce expectations so that they are less disappointing, and to increase effort step-by-step so that genuinely desired goals can be at least approximately reached.

FURTHER CLINICAL CONSIDERATIONS

Depression is uncomfortable, sometimes devastatingly so. It has an impact on the individual, but it can also have an impact on families and relationships, and even society. Reductions in depression can result in improvements in relationships, and improvements in relationships can help reduce depression. The next step is to look at relationships from an ops perspective.

Chapter 15

Relationships

Love and Work, Laughter and Dreams

Jane Austen, in her novel *Pride and Prejudice* (1813/2010), wrote perhaps the quintessential dramatic conflict between cognition and feelings. Elizabeth Bennet was one of several daughters in a proud but moderately poor household in Regency England. Her mother, Mrs. Bennet (in the role of cognition), wanted Elizabeth (in the role of feelings) to marry sensibly, someone with good prospects, the kind of person with a bit of land and a house, perhaps, whether Elizabeth truly loved him or not. Naturally, there is something to be said for the common sense of mothers. However, Elizabeth was firm in her resolve—she famously refused to marry someone she didn't truly love. Then the wealthy and well-connected Mr. Darcy began frequenting the area. However, Elizabeth was not sure she liked Mr. Darcy's laconic, superior aloofness. And yet, finally, she was not completely indisposed toward him, either. They met, they strolled, they bantered. Eventually, they warmed to each other. Eventually, they married.

She won on both criteria—love and gold, so to speak. However, Elizabeth Bennet is a fictional character, a person from the realm of imagination. The person from the realm of reality, the author, Jane Austen, never married. It is hard to know why at this point in history. Perhaps she simply wasn't interested. But it may be that Jane Austen herself was a little aloof, critical, and unrealistic (Lane, 1996). She lived with her family her entire life, in genteel, fragile poverty, never quite acquiring her full independence or a secure source of income, and died at an early age. She was absorbed, as a novelist, in a gracious but imagined Regency world where, "Wishes can come true, while understanding that many lives are lived in disappointment" (Austen, 1813/2010, p. 24). It may be that if Jane Austen herself had been more interested in finding and adjusting to her own Mr. Darcy in real-life *Pride and Prejudice* and Elizabeth Bennet might have taken on a very different character—a bit more cheerful and pragmatic. (Though literature might be the poorer for it.)

Leaping into the issue of relationships from a very different angle was D'Artagnan and *The Three Musketeers* (Dumas, 1844/2011). D'Artagnan was the classic poor boy from the country who ventured into the big city of Paris with dreams of joining the illustrious Musketeers, the King's elite guards. Three Musketeers—Athos, Porthos, and Aramis—almost immediately pegged D'Artagnan as a grandiose, upstart buffoon who was in want of a lesson, and all three Musketeers quickly engaged him in duels. To their amazement, D'Artagnan proved to be highly athletic and a superior swordsman who had little trouble defending himself. All was forgiven, and the four became fast friends in a blaze of literary glory.

The joy, finally, of the *Musketeers* is not the political intrigue, nor the fighting; it is the rollicking good *relationship* of the four Musketeers and the *personalities* that underlay that relationship—bouncy, warm, and enthusiastic, lighthearted yet serious, they are full of fearless bonhomie and rapier-quick repartee. And, perhaps instructively, their signature cry is—*one for all, and all for one.*

DEFINITION OF RELATIONSHIP

There are perhaps as many appealing definitions of *relationship* as there are pages in the collected works of Austen and Dumas. But for purposes of the ops system of personality, a *relationship* could be defined as *an exchange contract between two or more people that typically involves frequent contact.* That includes friendships, dating, marriage, family relationships, clubs, careers, diplomacy, and teams, such as at work or in sports.

That contractual part might sound strange, but, after all, a marriage is a relationship, and that is what it is called—a *marriage contract.* There is a presumed exchange in marriage that is a largely unspoken reality. The contractual foundation is not always made clear at the beginning, but may show up in court if the expected exchanges do not occur. It often shows up in conflicts within marriage, when unspoken expectations are not met. And, of course, *one for all, and all for one* is essentially a contract.

The contract part of a relationship is usually kept subliminal or unconscious, except in situations like law, business, or diplomacy where arrangements have to be explicit. The reason most relationship issues are kept nonconscious may have something do with the difference between the conscious and the nonconscious planes of the mind. The conscious plane seems more focused, exacting, practical, shrewder, and more calculating. The nonconscious plane seems more diffuse, ambiguous, imaginative, and, if a little lost, more romantic.

There seems to be a general feeling that as long as relationship arrangements are kept largely in the nonconscious plane, they are more likely to be sincere, from the heart, not calculated, and thus more trustworthy. So, it seems a little odd to define a relationship as an exchange contract—that seems so *conscious*—but upon examination that does appear to be the reality. Defining relationships that way can clarify their nature a great deal and can assist in identifying and resolving conflicts, and avoiding unpleasant surprises and presumed expectations that then go unmet. The Wolins (1993), regarding resilience, and Robin Norwood (1985), regarding decreasing hurt in partnerships, both suggested that a reasonable degree of explicit pragmatism in *thinking through* the costs and benefits of a potential relationship greatly aids in establishing satisfying, durable relationships.

THE FUNCTION OF RELATIONSHIPS: AN OPS PERSPECTIVE

To speak of the *function* of relationships almost sounds like an oxymoron, a *non sequitur*. Is not the whole *idea* of a relationship to plunge into a world of pure experience and to exit the world of pragmatic assessments? To a degree. But, from an ops point of view, relationships can be marvelous on their own, *and* they do have a function. New relationships tend to rapidly crystallize into commitments. Therefore, it might make sense to consider the rational side of relationships.

It might be easiest to do that by considering prototypical hunter-gatherers—a female, Abba, and a male, Bubba. Abba is all alone. She feels lonely as she walks the hills looking for food. She is vulnerable at night slumped up against a rock trying to get some sleep. If she is injured there will be no one to take care of her. She has no one for company and no one with whom to talk over difficulties, options, and plans. Survival is threatened.

Then she meets Bubba. Bubba is in the same situation as Abba—all alone, trying to figure out where the food and the water are, keeping an eye out for predators, fending for himself, one bad accident away from catastrophe. They realize their chances of survival are much, much better *together*. They link up. Pretty soon, they have children.

Pretty soon (to extend the story of this prototype couple), they live in a village. Now life is much more interesting, and much safer (barring the odd plague or tavern brawl). There are always friends and family to help and get help from, and always people to talk to and do things with. People incline toward specialization, and that makes life more profitable, more organized, and more efficient—the farmer and the grocer, the artist and the singer, the doctor, the midwife, the nurse and the nanny, the teacher, the tavern-keeper, and the soldier.

So, relationships, personal, social, and vocational, can be seen as having a lot to offer: stimulation, companionship, emotional support, amusement, specialized knowledge and skills, financial and physical security, cooperation in problem-solving, decision-making, task completion, and sharing of food and shelter, not to speak of assistance with raising children and surviving in injury or old age. A lot of that becomes *invisible* in the context of modern life. Because, as discussed previously, when contact is ubiquitous, people *habituate* to the benefits of relationships and *sensitize* to the costs. Then they may wonder, "Why am I going to all this trouble for this (or that) relationship?"

TYPICAL CLINICAL COMPLAINTS REGARDING RELATIONSHIPS

There seem to be four typical areas of concern in relationships—dating and marriage, and coworkers and supervisors. That perhaps distills down to Freud's famous brisk and categorical remark that a successful life consists of success in *love and work*.

Dating, clinically, seems to have two phases—what might be called the *courtship* phase and the *consolidation* phase. Usually there are not too many complaints about the courtship phase. People are on their best behavior then, focusing on strengths and pleasures, disguising or ignoring limitations and quirks. In the consolidation phase there seems to be a subliminal recognition that cards need to start being put on the table, partly as a test, to see if the real personality is acceptable, and partly as a prelude to a marriage, when all secrets, more or less, end. It is in the consolidation phase that complaints begin to arise. The basic clinical complaint is a mis-calibration, a mismatch between the expected template of the loved one and the eventual, apparent real template of the loved one. Thus, Person A would like to see Qualities 1-2-3-4 in Person B (that is the expected template). The initial impression was that Person B had all those qualities. But that was the courtship phase. The consolidation phase starts to reveal the real personality template. That often results in confrontation, negotiation, and adaptation, all of which can be difficult.

Similarly, marriage, when there are clinical difficulties, seems to have two phases—what might be called the *honeymoon* phase and the *awareness* phase. Each seems to last about two and a half years. Clinically, five years often seems to be sink-or-swim time. Complaints in the honeymoon phase are usually moderate—an irritating quirk here, an odd notion there, an unpleasant remark here, a questionable expenditure there. Most people subliminally expect in marriage to learn some new things about their spouse, and to have to make some adjustments as they form a unified system that can accomplish the tasks of a married couple and, usually, the tasks of parents. However, if negative off-template (undesirable) attitudes and behaviors persist and seriously undermine the balance and functioning (homeostasis and output) of the marital system, then serious questions may be raised in the awareness phase about the viability of the relationship. To maintain a marriage, somehow the "Core relationship created out of the conscious and unconscious fit of the partners' need and wishes" needs to be maintained (Wallerstein & Blakeslee, 1995, pp. 20–21).

Work complaints about colleagues seem to be something like dating complaints—irritations and conflicts over habits and attitudes, pretensions and presumptions, which, to some extent, come and go. After all, people can always close their doors, choose someone else to go to lunch with, or even ask for a reassignment, just as people can find someone new to date.

Supervisors are a little more like spouses—a major relationship linked to money, self-esteem, effort, time, and critical future prospects in which there is only modest room for major negotiation and which can be discontinued only at great cost. With all that pressure, it is no wonder that tempers can boil. The usual complaints about supervisors are that they are too demanding and critical; they are cold, unsympathetic, and inflexible; and they are only concerned about the bottom line. But, interestingly, it has been cogently argued that a supervisor often has to take on that firm command role, with clear-eyed attention to profit, and thus working under that brisk, confident supervisory style is one of those work realities to which supervisees need to adapt (Hill, 2003; Sandler & Gray, 1999).

PORTRAIT OF HIGH-FUNCTIONING OPS PERSONALITY IN RELATION TO RELATIONSHIPS

The essence of a high-ops person is a confident mastery of self and responsibilities, of the individual *career* of living. A working assumption is that other people are similarly inclined and similarly capable. That means that, typically, in building a relationship *two capable and responsible people are allying*. That means that they can understand each other, can trust each other, can collaborate on identifying goals and means, and can have fun in the process of *operating together*.

That may sound like a lot of trouble, somehow. But it almost always works that way, even when there is almost no actual work involved. For example, someone may ask someone else out on a date, or for lunch, or to the beach, or to go shopping. Those are all enjoyable activities, but they are activities, with some kind of purpose and characteristic aspects. They are all *operations*. *Relationships are almost always premised on operations—shared activities leading to shared goals.* If people don't share goals much, there probably isn't going to be much of a relationship.

People who are high functioning in relationships tend to take relationships seriously—relationships are important to them, so they put effort into establishing and maintaining them. They accept that a relationship is a shared, cooperative arrangement that involves giving as well as receiving, and frequent discussion, negotiation, and compromise. They make sure they are caring and responsive partners (personal) and team members (sports, work). They are comfortable clarifying what they are looking for in a relationship—what they hope to gain and what they are willing to give in return. They keep an eye on their relationships to make sure they are working. If there are difficulties they like to politely but reasonably quickly get those straightened out. Because they prize good relationships, and because they know they need and benefit from good relationships, and because they put time, energy, money, and hope into their relationships, they usually premise their relationships on reliability as well as interest and creativity—they prefer to establish friends, marriage, and career for the *long term.* They genuinely grieve when relationships end. They even hate to see the neighbors move away (except for those people who never returned the stainless steel clippers!).

RECALIBRATION OF OPS COMPONENTS: RELATIONSHIPS MAKEOVER

- *Ideals:* The first step in strengthening relationships is reflecting on the *value* of good relationships. Some people would propose that good relationships are the essence of a good life. Many people have been surprised to find that at the end of a long and materially successful career, what is really important is not those material accomplishments, as useful and reassuring as those might be, but rather, a good marriage, good friends, and a good relationship with children and grandchildren.

  Often the best things that happen in life are those radiant moments of a really good relationship. And, often the worst things that happen are those dark moments when a relationship turned toxic. A recognition of that human reality may help to recognize that relationships can be so important that they justify the time and effort that go into good ones. At the very least, the process of *evaluating* the ultimate worth of relationships is probably a useful investment in time.
- *Cognition:* Again, there is a certain, usually unspoken give-to-get premise in establishing a relationship, a set of presumptions as to what will be gained (Satir, 1967). Characteristics that could be exchanged, such as sense of humor or athleticism, can be conceptualized as occurring on a 10-point

scale. The default evaluation would be 5. A low score would be a 1 or a 2. A high score would be a 9 or a 10. An average prototype match would be two 5's connecting.

For example, in this psychometric scheme, in relation to sense of humor, two people both scoring 5 could be considered to each have a reasonably adequate sense of humor. A 1–2 would be someone rather serious, hard to laugh with. A 9–10 would be someone witty and amusing, lively company. In relation to athleticism, a 1–2 would be someone disinclined toward physical activity, a reader or an artist, perhaps. A 9–10 would be someone seriously involved in fitness and sports, probably quite fit, vaguely resembling a superhero. There are usually some *tradeoffs* regarding personal characteristics. Thus, someone not so amusing could *substitute*, in the exchange, striking physical condition for amusement.

As an example of this tradeoff from the movies, Annie (Diane Keaton), in *Annie Hall* (Greenhut & Allen, 1977), has a striking appearance (perhaps a 9–10), though she is not so witty (2–3). She is matched with Alvy (Woody Allen), who is not so impressive physically (perhaps a 2–3, as he portrays himself), but has a quick and entrancing wit (9–10). Thus, they are attracted to each other, and the relationship is somewhat balanced in its asymmetrical way, but it never quite gels, and finally ends in regretful confusion.

One way of looking at that failed relationship is that the tradeoffs (the asymmetry) between Annie and Alvy were *too* extreme; that is, there was a certain exchange balance, but then there wasn't enough *functional commonality* to form a viable *relationship system.* There isn't going to be a perfect match, usually, but a rough quantitative ops guideline might be: no more than four points apart. That is, some complementarity (a strength in one partner compensating for a limitation in the other) is alright, but symmetry (a good match in characteristics, strength-to-strength and limitation-to-limitation) tends to result in more emotional equanimity and understanding in the relationship. Complementarity tends to lead to *one-up and one-down* dominance conflicts (Watzlawick, Bavelas, & Jackson, 1967). "Why can't you ever . . ." being something of a signature complaint in complementary relationships.

- *Feelings: Gut feelings* (i.e., emotional intuition) are often a helpful bottom-line guide to relationships. Ideals can paint a beautiful picture, and cognition can help calculate the probability that a relationship will work, but if the relationship doesn't *feel* right, it is an indication that it won't work.

For example, someone could consider getting a job at a big organization with lots of perks and pay and opportunities. Friends and teachers might roundly endorse the choice. But if a couple of interviews and tours around the offices don't feel right, there may be something about the potential relationship that is registering subliminally as somehow *off*. On the other hand, if the *chemistry* (the underlying simpatico feeling) is good, that can balance out some apparent limitations.

- *Decisions:* There is a certain discipline involved in decision-making regarding relationships. Relationships are so important, and are often accompanied by such powerful hormonal surges that it can be difficult to actually *make a choice* as opposed to being *plunged into a situation.*

  Checking with friends and family members—who are usually not quite so caught up in the excitement—can help provide rational grounding. *Pro-con checklists* can help compare evident costs and benefits. *Forecasting and scenarios* can help provide useful probabilities as well as phenomenology for a client—the imagined experience of being in the likely relationship six months, two years, five years down the road. For example, on paper, marrying a doctor might look like a winner, but running a scenario on the relationship and imagining the probability of eating dinner alone many nights while the doctor-spouse is at the hospital seeing patients might make the relationship look and feel a little different.
- *Actions:* Overall, relationships can be seen as having three phases, each calling for different kinds of actions. The first might be called the *acquaintance* phase. Despite internal and external pressures to the contrary, and despite what goes on in movies or television where there is limited time to make things happen, it is usually neither necessary nor advisable to jump into a relationship. In real life, clients usually can benefit from *cultivating a centered, confident, evaluative frame of mind.* That does not have to be unfriendly. Just self-respecting and sensible. Not everyone who is thoughtful and fascinating in the short run turns out to be that way in the long run (e.g., Norwood, 1985).

  The second phase might be called the *friendship* phase. Multiple contacts have occurred in a variety of settings. Each has gone fairly well. The relationship seems reasonably balanced, revolving around lots of interesting and useful commonalities. Trust rises. Candor increases. A genuine sense of security and well-being materializes. The relationship is reliable and fun. A friend has been made.

  The third phase might be called the *soul-mate* or *best-buddy* phase. This is probably the gold standard for most clients. At this point there is tremendous trust and security; the relationship feels rock-solid. The people know each other really well and can fairly predict what the other person will do in any given situation. They are highly committed to each other and would literally risk their lives and fortune for each other. They genuinely enjoy each other's company whatever they are doing, even if they are doing not much of anything. Knowing these three phases can help guide the actions that lead to each, from casual pleasantries, to frequently shared and enjoyed activities, to reflexive trust and cooperation.

- *Objectives:* As is so often the case with operations, almost the hardest part of an ops approach to relationships is *thinking specifically*: "Where do I want this relationship to go?" "What kind of relationship do I want to have?" "By what criteria will I see that I have that—how will I know when I get there?" "What am I willing to offer, and what do I hope to receive in exchange?" Again, it is to some extent that struggle between conscious thinking and nonconscious thinking. It just almost doesn't seem *right* to many people to think specifically about what they *want*. But clinical experience suggests that clients who can cross that bridge from nonconscious to conscious, from vague to specific, will be in a far better position to establish relationships that are exciting and interesting, but also are likely to be satisfying and enduring.
- *Communication:* Feedforward in regard to relationships in the ops approach has, perhaps, two main parts. First part is sheer, consistent courtesy, something more complex than it appears at first (Post, 2014). The second part is self-effacing, amused candor; "Anything you say that is funny will, to some degree, have a hidden truth behind it" (Macks, 2003, p. 19), including people's jokes about themselves. It is hard to dislike someone who is polite. And it is hard to dislike someone who is cheerfully frank and funny. The two together create a presentation both *appealing and informative*, a promising basis for a relationship.

  Feedback is the tough one. It is easy to dismiss rejection from someone uninteresting. But rejection that comes from someone highly appealing, someone who launches a thousand expectations of love and laughter, of wealth and wisdom, can be painful, for example, Mr. Darcy getting turned down by Elizabeth Bennet in *Pride and Prejudice* (Austen, 1813/2010). The trick is to *not let expectations run way out ahead of fact*, to keep them harnessed together, like a team of horses and a carriage. If the person seems interested, then it makes sense to encourage the expectations. If the other person is a little cool and vague, then it probably makes sense to rein in the expectations. Then any rejection is not so surprising and not so painful. As CBT has pointed out, reality is much more manageable when expectations are derived from observations, from track records (e.g., Alford & Beck, 1997; Ellis & Harper, 1975).

SUMMARY

There are all kinds of relationships, from work and diplomacy, to sports teams and parenting, to, of course, romance. There seem to be two core elements to relationships—*contractual exchange* (the more pragmatic part) and *emotions* (the less pragmatic part). The emotional part is the more dramatic, and often seems to be in charge. But it may be that satisfying emotions in relationships are actually, to a surprising degree, based on making sure that the underlying deal, the contractual relationship, is fair, sensible, and reliable.

FURTHER CLINICAL CONSIDERATIONS

Relationships are often the prelude to marriage and children. Children may seem, in some ways, like an interruption in relationships. But they are really a whole new and primary set of relationships that will certainly last a lifetime. Many of the principles that apply to the personality and to relationships as systems of operations also apply to the family as a system of operations.

Chapter 16

Parenting

Building Happy, Healthy Families

A long time ago, according to a wonderful novel, *Swiss Family Robinson* (Wyss, 1812/2006), an honest, decent, hardworking Swiss family was caught in a storm, which crashed their sailing ship onto the shore of a distant tropical island. The family included a mother, a father, and four boys. They might have frozen with terror, but they did not; and they might have given up hope, but they did not; and they might have sat in their boat waiting to be rescued, but they did not. What they did was bravely launch into the middle of the situation and take charge. They cobbled together a crude small vessel, then they ferried animals and tools, food and weapons from the wreck to the shore. Then they planted seeds, they hunted and fished, and they built a marvelous and clever tree house, with running water. (That same tree house can be seen today at Disneyland; well, it is probably not the *exact* same tree house!)

Swiss Family Robinson amounts to an inspiring handbook on how to survive and thrive under frightening circumstances (and reality is sometimes like that). *Robinson* is quintessential Enlightenment thinking—how to apply science and organization to problems within a framework of confidence, energy, and purpose (e.g., Hampson, 1968). As the mother said, "Courage, then!, pursued I; my husband and my eldest son are at this moment employed for the general good; why should not I be active and enterprising also?" (Wyss, 1812/2006, p. 46).

There may have been a multiplicity of situations that would have illustrated that bold and rational style of thinking, but *camping*—which is what *Robinson* functionally comes down to—is a particularly good way to illustrate how high-functioning families operate. Robert Beavers, the well-funded, epochal family researcher (1977), mentioned that high-functioning (*optimal*) families have a somewhat odd penchant for camping. The reason, it turned out, is that camping is just about perfect for teaching how life at its most basic works; for teaching a direct connection between need, effort, and outcome; and thus for teaching responsibility and a strong sense of independence and competence.

The world appears highly complex, of course; yet an astonishing proportion of that complexity seems to flow from the human mind itself! Zen, in its penetrating brilliance (e.g., Paul, 2000; Suzuki, 1970), tries to reduce living to an aesthetic, functional minimum in order to get people in direct, serene contact with reality, so they can better understand and master reality.

Along rather less militant lines, the vacation destination of the island of Santorini in the Aegean seems to try to get visitors to drop the stressful complexity of their ordinary lives and experience the simplicity of classic Cycladic island living—red bougainvilleas and geraniums, simple white buildings, local wines, fresh fish from the surrounding sea, fresh bread, music and laughter, and friendly locals who, on the whole, love life and people. Those approaches—camping, Zen, and Santorini—are conscious and ultimately concrete, straightforward attempts to distill life to a functional essence, and so to enhance mastery and enjoyment of life.

It is not too surprising that in the often politically charged kaleidoscope of child-rearing and character development various thinkers have arrived at conclusions that *seem* to bring matters down to such solid and harmonious basics, but perhaps do not. For example, the philosopher Rousseau's treatise on child-rearing and education, *Emile* (1762/1993), appears to be an attempt to get people (i.e., his contemporary culture) disentangled from the kind of absurd complexities of stiff propriety and formulaic relationships alluded to in Jane Austen's work (1813/2010). Rousseau was making an attempt to bring people to a noble naturalness. The educational procedure for doing that was to let the fictional protagonist Emile do whatever he wanted! Rousseau himself was successful as a writer and a revolutionary theoretician, but unsuccessful as a tutor and a complete failure as a father (Wokler, 1995). His pedagogical approach proposed for the fictional Emile would appear to be something like *chaotic overindulgence*, an approach to character development that, clinically, often seems associated with poor self-regulation, impulsivity, and AD/HD, among other difficulties. For example, "The only habits a child should be allowed to contract is that of having no habits" (Rousseau, 1762/1993, p. 34), "Nature provides for the child's growth in her own fashion, and this should never be thwarted," and "Let them run, jump, and shout to their heart's content" (Rousseau, 1762/1993, p. 58).

John Dewey was a very prominent educator, philosopher, and psychologist of the early twentieth century, the founder of *progressive education*, and also a devotee of Rousseau. Dewey appeared to be trying to free the education of children from the overscrupulous, overcontrolling upbringings of the Victorian and Edwardian eras. But, like Rousseau, he seems to have erred on the side of *laissez-faire*, and to also have been inclined toward an uncertain metaphysical perspective that may have distorted his thinking. For example, "I believe that the school is primarily a social institution"; "Education, therefore, is a process of living and not a preparation for future living" (Dewey, 1998, p. 230); and "The teacher is not in the school to impose certain ideas or to form certain habits in the child" (Dewey, 1998, p. 231). It is hard to see in such statements an actual educational process; they sound more like a description of an after-school enrichment program. They don't sound like they are aimed at giving children a substantial platform of information, intellectual skills, and organizational habits that would prepare them for success in a competitive world.

Likewise, Everett Shostrom (1968), a renowned and influential psychologist from the mid-twentieth century, seemed to be trying to free people from the stifling conformity and rigidity of the 1940s and 1950s. His primary suggestion was to exit overcontrol, then to be open to experience, to be self-accepting, to be humane, and to let the personality evolve pretty much on its own terms, at home and in school. Not all bad suggestions, but not quite a handbook for competitive success, either—not quite providing adequate direction or actual means of succeeding in life. Shostrom's thinking, too, may have been infected by a vague metaphysics.

These three thinkers—Rousseau, Dewey, and Shostrom—together seem to have had an enormous laissez-faire influence on modern thought about education, child-rearing, and socialization. Some of it seems helpful, but mostly it seems counterproductive, rather a variety of directionless, *postmodern* (unstructured and purposeless)-style educating and parenting (e.g., Derrida, 1978; Silverman & Ihde, 1985; Skinner, 1985).

DEFINITION OF PARENTING

Thus, it could be argued that, to be effective, parenting must be based on a viable *theory of living* that more or less corresponds to reality, something beautiful, creative, and enjoyable, perhaps, but still solid—well-oriented toward managing a successful life. That, as the brief review above suggests, is harder to do than it looks. Perhaps an attempt at a paradigmatic definition will be useful: *parenting is the process of showering affection and care on children while teaching them the principles of successful living, developing in them the character to fulfill those principles, in a fun and creative yet responsible atmosphere that allows them to develop a unique personality.*

THE FUNCTION OF PARENTING: AN OPS PERSPECTIVE

The ops view of parenting is similar to the brilliant model presented in *Swiss Family Robinson* (Wyss, 1812/2006) and similar to the research-based model of optimal families (Beavers, 1977). Life presents certain demands to the individual. To meet those demands, to survive, the individual must have relevant knowledge and skills, must be good at establishing and maintaining relationships with others, must bring energy and courage to bear on the problems of living, and must be reasonably well organized and prudent, and all that with at least a modicum of fun, even with flair, and a great deal of satisfaction if the inevitable disappointments and difficulties are to be surmounted.

Infants, naturally, know little of all this. They know only that they are hungry or full, too hot or too cold, lonely or comforted. It is up to the parents to *create the environment* in which children can thrive and, in their own way, prosper. And the first, prior, step in creating that environment is perhaps in the parents sketching out, in their view, *a good life*. More than a little thought has gone into this topic so there are many sources of ideas and inspiration, for example, *The Good Life* (Cicero, 1971), *The Happiness Hypothesis* (Haidt, 2006), and *Happy* (Smith, 2010). Positive psychology itself, as a field, has numerous suggestions as to how to live well (e.g., Fredrickson, 2009). That parental sketch then becomes, in general, the *idea* of what their home will be like, their philosophy—how their home will be run, what their children will be like within broad boundaries, and, to some extent, the *meaning* of their family.

People, including parents, exist within a *matrix of relationships* (Ruesch & Bateson, 1951). Therefore, the parents have a certain *contract* with society—an unspoken commitment to the neighbors—to raise children who will be reasonably productive, helpful, thoughtful, and happy (Mill, 1863/1987). *Good kids.* (Would questions not be raised about parents planning on *not* raising good kids?) And parents have a certain unspoken contract with their children—to guide them in directions that probably will lead to success and satisfaction in later life, directions that parents know but children do not know. Homework, good nutrition, exercise, and a reasonably positive attitude are traditional areas of conflict. But that is part of the function of parenting—to make sure that children master those challenges so they will have effective habits later in life (Beavers, 1977; Covey, 2004). That is, it seems that there are certain recurring *verities* to human life and society that have to be successfully dealt with if serious difficulties are to be avoided. Those are, in a sense, the *formal parenting goals and tasks*. These are primarily the issues of *character*.

Different than character development, but still interesting and important in human life, are the idiosyncratic and cultural traditions of a family, the *personality* of the family. For example, a family with a German background might make a big occasion of Oktoberfest, and a family with Irish roots might make a point of cheerfully honoring St. Patrick's Day. Some families take sports very seriously, others music. These individualistic touches are part of the uniqueness and closeness of family life, part of what makes family life so wonderful.

Last but not least are the *temperaments and personal tastes of the children*. All children are different. Some are *by nature* artistic or athletic, social or contemplative, poetic or competitive, and so on. *Chance* also plays a part. For example, a charismatic elementary teacher might fire up an interest in art or music, theater or science; a gung-ho high school coach might inspire a love of athletics; or a nice neighbor might spark an interest in a law career. Nature and chance then combine to create unique combinations of interests and ambitions. If those personal proclivities are ignored or squelched, that seems to have a negative, suppressing effect on children's enthusiasm and love of life, as well as the breadth and depth of their personalities. By contrast, children whose talents and interests are encouraged seem to blossom in interesting and impressive ways. This is suggested in studies of genius (e.g., Robinson, 2010; Simonton, 1994).

Thus, the function of parenting from an ops point of view is to raise human beings who are confident and capable, creative and productive, who get along well with others and function well within systems in which they have a stake (like schools and teams), and who are reasonably cheerful themselves and respectful toward their parents. Such children—well grounded in social, behavioral, and intellectual skills—are *manifestly* on their way to being happy and successful adults. It is up to the parents to, in a sense, *design* this family system into which the children will be born, and then to manage the system so that everyone succeeds. The concept of *psychosocial design* is somewhat similar to the concept of *engineering design* (e.g., Petroski, 1996). This sounds complicated, but many parents, of course, find it comes naturally.

The desirable personal characteristics that are, in general, the goal of child-rearing are captured in the ops model in the seven personality components. Thus, it is those components that become the focus of parenting in the ops system.

TYPICAL CLINICAL COMPLAINTS REGARDING PARENTING

Clinically, the usual complaint of parents is that their children do not behave properly. The content of that behavior includes keeping rooms clean, doing homework, helping out around the house, being respectful to parents, and getting along in school. Clinically, the usual complaint of children is that they don't have enough pleasant, productive contact with their parents, and their parents don't quite seem interested in their personalities or their lives.

PORTRAIT OF HIGH-FUNCTIONING OPS PERSONALITY IN RELATION TO PARENTING

High-functioning parents enjoy their children and have fun with them (Beavers, 1977). Nonetheless, they also hold their children responsible for participating productively in the family, for maintaining a good attitude, for being respectful and reasonably compliant, for thinking for themselves, and for taking good care of themselves. They encourage their children's interests and talents, their friendships and their activities. They often talk to their children in a deep and serious way, genuinely connecting with their children's thoughts and feelings, hopes and dreams. The children feel cared for and appreciated, but also know they are responsible for functioning adequately. On the whole, the family has fun and gets things done together and as individuals; they are successful and functional, and they also enjoy each other's company. They savor life in general and family life in particular.

RECALIBRATION OF OPS COMPONENTS: PARENTING MAKEOVER

As people often say—children don't come with an instruction manual. But, really, there are parenting suggestions all around. One source of guidance is the kinds of books that have been loved by generations of parents and children. Books for young children are often cozy and sweet, kind and thoughtful, cute and reassuring, like *Winnie-the-Pooh* (Milne, 1926). In that vein, elementary schools are often full of pictures of rotund panda bears and koala bears, butterflies and sunshine—symbols of warmth and safety, love and attachment. The intent is to *lay a foundation of confidence and security, an emotional refuge and platform, positive bonding* from which later adventures can be launched, such as going off to college, going off to a new job, traveling around the earth, or even landing on the moon. Just so, children's stories quickly progress to adventures and excitement, and indirectly provide instructions in coping with danger and wild confusion, like *Peter Pan* (Barrie, 1995) and *Alice's Adventures in Wonderland* (Carroll, 1946). As children advance in years, much-beloved classic fiction stories begin to address more directly the dangers of the world and the need to balance shrewdness with risk, such as the plucky Rover Boys (Stratemeyer, 2007), the brave and clever Nancy Drew (Keene, 1957), and the noble Diana Prince—Wonder

Woman—created by a psychologist to inspire girls (Daniels, 2000), to cite some traditional American examples.

In other words, parents are surrounded by good models, in literature or in real life—perhaps a happy and productive family down the street or cherished relatives. It is partly a matter of looking for appealing and successful models, and then personally imitating good parental models, as well as encouraging children to imitate capable child and adult models. And what makes a good model is partly a matter of the experience and values of the parents. And, of course, a lot of it is *love and fun*—that intuitive sense of caring and connection, and enjoying each other's company.

- *Ideals:* Strangely, or not, children rapidly develop a sense of a good world, good parents, good friends, a good life. Ironically, clinical experience suggests that perhaps the sharpest sense of decency and goodness comes from young people with marked difficulties, such as being incarcerated in juvenile hall, possibly because they see such a clear contrast between their own homes and that of children who do not get into serious trouble. Among all children, those in serious trouble seem to yearn most acutely for a just world.

  Human nature can be marvelously kind and caring, warm and fun, such as when gathered around the hearth in the evening, eating, chatting, perhaps singing songs together, but, by necessity, people can also be hard and cold, such as when defending their prototypical apple orchard. So, shrewd toughness definitely has its place in the scheme of things. Nevertheless, on the whole, what people like in other people is kindness and thoughtfulness, hard work and responsibility. Those decent and idealistic people, it turns out, are the people who head great organizations (Collins, 2001), thereby contributing to society as well as making handsome salaries. Therefore, it is in the interests of children to *develop ideals and values on the high side of life*. Being human, children, especially young ones, have a hard time sorting out the not-so-great things that feel good (like taking) from the great things that don't always feel so good (like giving). It is up to the parents to steer the children away from the Darth Vader side of life, so to speak, and build into children a strong intrinsic motivation for doing the right thing even when that is hard. That is, it is up to parents, at least initially, to develop that *Star Wars* Force (Lucas, 1975) in their children, which largely comes down to *the force of character—noble ideals blended with pragmatic values consistently followed!*
- *Cognition:* Rousseau (1762/1993), Dewey (1998), and Shostrom (1968) are nothing if not good writers. It is easy to be carried away by their glamorous and mysterious implication, overt and covert, that almost the only thing necessary to create exemplary human beings is to do nothing! Would that

it were so. If parents just wouldn't *interfere* with children's development, the earth would be awash with noble beings! Not quite, of course.

Partly, a successful life comes from being an effective general problem-solver (Garnham & Oakhill, 1994). As the distinguished psychiatric researcher Jerome Frank said (1974, p. 330), in the process of therapy, the patient "Becomes able to tackle the problems he had been avoiding and to experiment with new, better ways of handling them." Effective problem-solving depends on the development of numerous cognitive elements: calm alertness, attention control, defining terms, being categorical and logical, quantifying, and basing predictions on facts. In other words, not too different than what goes on in the average elementary school classroom. Parents can help build up cognitive strengths in their children by *expecting* them to think well (prompting) and then by *reinforcing* good thinking by saying things like, "Good observation!" or "That's very sensible and logical." In that process, children's *high intellectual competence will be linked to high self-esteem.* Most usefully, parents can model clear and rational thinking, forthrightly addressing issues in a factual way.

- *Feelings:* Sensations and emotions are, in a way, primary, as indicated previously—what is more important than being hugged and kissed and loved? *Affection and bonding are a basic part of being a family.* And, naturally, at least traditionally, in their hearts—every father believes his daughter is the most beautiful, accomplished girl in the world, and every mother believes her son is the handsomest, most capable boy in the world.

  And yet, on a day-to-day basis, feelings are probably secondary to task accomplishment! That is, in a way, basic reality: no paycheck, no dinner; no homework, no diploma. That give-to-get, economic exchange arrangement in life (as mirrored in the mind's *picoeconomics*) makes parenting a little easier. What parents can look for is *focused activity* (even if that activity is creative) accompanied by largely positive emotions. Roughly, if the activity is successful, the emotions are positive; by contrast, if the activity is a wash or not successful, the emotions are neutral or negative. This is a very *basic linking of emotion to outcome*, the logic of which, strange as it may seem, is not always obvious to children. Parents *can* help their children by emphasizing this do-good-to-feel-good logic. This teaching provides intrinsic motivation for children to be successful, a mentality that will assist them wherever they go. This exchange principle is a good follow-up to the more primary feel-good-because-you-are-loved family bonding.
- *Decisions:* As with the pattern of changes in the solar system (to which humans seem closely linked) and good sleep patterns, for a family system to function, it is usually better to have *regularities*—such as mealtime, fun time, homework time, and bedtime. But in the family, as in any decision-making system, those regularities are expressed as *rules*. The ambiguity

and sometimes arbitrariness of rules, combined with the sometimes contrariness of human nature, often results in push-back from the children—rebellion, or at least testing the limits.

What to do? Successful parents (Beavers, 1977; Bodenhamer, 1983) seem to have a three-part approach. (1) They *assert the right to make the rules*; after all, they are paying for everything. They make those rules crisp and clear, and they *stick to them.* No wiggling, no excuses (usually). (2) Nonetheless, parents go out of their way to explain their reasoning and to get the children's input. They seriously *consider the children's views* and accommodate them when that is sensible. (3) As children mature, the parents *progressively transfer* the decision-making responsibility to the children as the children show they can handle it, while parents *retain veto power.* (The parents are still paying the mortgage!)

- *Actions:* There is an old conundrum in parenting, and living generally, about *being* versus *doing* (e.g., Husserl, 1964; Maslow, 1968). But the apparent contradiction may be another paradox—as long as children are *loved,* it is perhaps reasonable to expect them to be *doing something constructive* most of the time, like reading or pursuing a hobby. That may sound onerous, but passivity seems to be something like opium—a little goes a long way, it is easy to get addicted, and it is hard to recover; it may even negatively affect brain development, in that abilities not developed tend to atrophy (Nelson, de Haan, & Thomas, 2006). Clinically, passivity appears to be a fertile field for anxiety, depression, and restless resentment, as well as general failure in the tasks of living. Focused activity is more a mark of childhood health.
- *Objectives:* Goals are like torches at the front of a line—they provide direction and a sense of purpose. Goal setting is another one of those things that children may never quite fully stumble on themselves. And it is another one of those things parents can *easily* teach merely by talking it up. "What do you like?," "What do you want?," "What would make you happy?," and "How could you get that?" are examples of operational questions that might never occur to children but that can be easily asked by parents. Once children grasp the way that *goals connect the inner, phenomenological realm with the outer, objective world,* they come alive to the whole idea of having and pursuing dreams. Relevant activity is then a natural and prized outgrowth of goal setting. The marvelousness of using ambition and goal setting to escape poverty and trauma can be seen in the touching autobiography of the best-selling singer Nana Mouskouri (2007), who, through enormous effort and her family's encouragement, lifted herself out from desperately war-torn, cold, and hungry circumstances to achieve worldwide stardom and wealth, with her compassionate and generous values intact.

- *Communication:* Sharing information and affection within a family probably has an almost infinite number of individualized shades and nuances. But perhaps it is not so complicated.

  Most parents could probably pick out of a crowd of families shopping and dining, say, down at the local mall, the parent-child communication style they would like to see in their own homes. Based on clinical complaints and successes, that prototype style profile might go something like this: warm, clear, honest but tactful and respectful, interesting, enjoyable and perhaps even amusing, thoughtful, and focused on facts and events but drawing on feelings and wishes—enhancing close family ties while sharing an interesting activity.

  Actually *building* that style of communication into parenting is not all that hard. It is a matter of consistently *modeling* that style of communication, *encouraging* that style in children, even explicitly requesting it, and then periodically *praising* it (intermittent reinforcement). (And if the children are *really* well-spoken and delightful, maybe even stopping off for a few fresh-baked, chocolate-chip cookies, or some other wicked treat.)

SUMMARY

The first principle of parenting is to love children and have fun with them. Engendering in children a profound sense of affectionate belonging and physical and emotional security is foundational. However, that having been established, it is critical that children learn to function effectively. That means parent-guided character development and goal-oriented activity. Thus, necessary knowledge, skills, and attitudes can be honed to that competitive edge that will allow children to succeed in the challenges of life—and be happy.

FURTHER CLINICAL CONSIDERATIONS

A shocking thought is that the AD/HD epidemic in America partly may be due to what seems to have been a national shift in parenting and educating style from structured and focused to laissez-faire and progressive. The ops approach, with its emphasis on goal-seeking, is well-suited to dealing with attention-deficit tendencies.

Chapter 17

Attention-Deficit/Hyperactivity Disorder (AD/HD)

Shattered Operations

"Nirvana Don't Come Easy," to paraphrase the title of a song written by the much-loved Beatles drummer, Ringo Starr. Nirvana is a famous state of mind in which everything falls into place, in which there is no disconnect between the way things ought to be and the way things seem to be—a state of serene but alert acceptance (Johansson, 1969). That may sound easy, but thousands of words have been written about how to achieve that state, and untold hours have been spent in ashrams, spas, and karate dojos across time and around the world trying to achieve that state. Matthew Polly (2007), from Princeton University, discovered that the glamorous kung fu training at the Shaolin Temple in China required an unbelievable number of repetitions of small moves to achieve apparently effortless expertise and a consequent state of smooth, confident control. The psychologist and Green-Beret trainer Richard Strozzi-Heckler (2007) has a whole institute (dojo) that teaches martial arts and meditation aimed at developing controlled, resourceful mental focus. Asian culture appears to have something to teach the West about mental focus that bears on psychotherapy.

Asians, as a group (Nisbett, 2003), appear to be very comfortable with the Confucian idea of refining calm attention control and then linking that to patient repetition in an almost *n*th-degree learning approach that results in mastery over intellectual and physical systems—something like a meditational mindset applied to problem-solving. The Programme for International Student Assessment (PISA) provides an objective comparative evaluation of the performance of fifteen-year-olds worldwide in mathematics, science, and reading. A BBC report (British Broadcasting Corporation, 2015) indicated that at the top of the mathematics food chain were Shanghai, Singapore, Hong Kong, Taiwan, South Korea, Macao, Japan, and Liechtenstein.

(Thank goodness for tiny Liechtenstein, or the West wouldn't have cracked the top *eight*! Talk about being out of medal contention—the United States was thirty-sixth, should anyone ask. Gold, silver, bronze, and *aluminum foil!*)

It is hard to identify exactly the sources of this stunning East-West discrepancy in performance scores. One approach (something of a loose, conjectural field study) would be to look at what kinds of *cultural changes* have occurred in the United States in the last fifty years, and try to determine if there is a loose correspondence between changes in the culture and changes in intellectual functioning, including attention control, that might result in such a decrement. And, indeed, there does plausibly appear to have been a shift in national attitude from what might be thought of as *structure*, including attention control, to *nonstructure*, including routine, intentional deviation from careful focus and tracking, which is to say attention-deficit habits.

For example, in the 1950s in the West there seemed to be a strong emphasis on being *well-adjusted*—on *adapting* the operations of the mind to school and life (Piaget & Inhelder, 1969), on having a *strong ego* capable of adapting (Hartmann, 1964), on being *competent* (White, 1959), and on *achieving* a lot (McClelland, 1967). That seemed to be the culture—*focused, ambitious, capable, cooperative*. Then came Woodstock.

The 1960s and 1970s were explicitly a *counter*-culture (Roszak, 1969), a *repudiation* of many of those prior principles of conscious attention and achievement. That revolution was enshrined, perhaps, in those charismatic, emblematic lines from "Mr. Tambourine Man": "There is no place I'm going . . . the jingle jangle morning . . . left me blindly here to stand . . . I have no one to meet . . . my senses have been stripped . . . yes, to dance beneath the diamond sky with one hand waving free . . ." (Dylan, 1964).

Of course, as mentioned earlier there *were* psychosocial difficulties in the 1950s and the early 1960s (Perls, Hefferline, & Goodman, 1951; Schaar, 1961). There was an alarming amount of personality rigidity. And there was pervasive guilt over failings great and small, past and present, actual and conceivable. And there was a consuming anxiety about the reactions and opinions of other people, captured in the fretful lines from the 1950s play for voices *Under Milk Wood*, "Oh, what'll the neighbours [*sic*] say, what'll the neighbours . . ." (Thomas, 1954, p. 10).

But all that energized, socially oriented, technologically sound, focused ambition did indeed accomplish a lot—winning World War II, building a prosperous society, laying down the interstate highways, landing Americans on the moon, and so forth. All that structure, energetic compliance, and respect for reasonable authority did have its payoffs.

It seems like a shame, and in hindsight was perhaps unwise, to have thrown out the baby (successful planning and production) with the bath (excessive rigidity and alienation). Even today a certain unstructured, directionless approach to child-rearing, in which experience and individual expression are almost everything, in which structured functioning based on systematic attention control seems to be given short shrift, appears to be alarming no less than the breezy, cosmopolitan *Time* magazine (Steinmetz, 2015). Currently, some schools in the south diagnose fully 25 percent of their students as having AH/HD (Leo & Lacasse, 2015). The Centers for Disease Control (n.d., p. 2) have reported that, "Rates of ADHD diagnosis increased an average of 3 percent per year from 1997 to 2006 and an average of approximately 5 percent per year from 2003 to 2011." That is a very large, progressive, and mysterious increase. If it were cholera, the overpopulation of the United States would be a thing of the past. Indeed, there may be a public health/mental health parallel in that the biological contamination in water, as with cholera, is largely invisible, just as a possible ideational contamination, so to speak, of the culture could exist largely outside societal insight.

DEFINITION OF ATTENTION, ATTENTION CONTROL, AND ATTENTION DEFICIT

Within the ops system, *attention* could be defined as *perceptual and cognitive focus on a target. Attention control* could be defined as *the capacity to shift focus on or off a target and to sustain focus despite distractions and fatigue.* This is the kind of process detailed in *Zen Meditation Therapy*—Zazen mental control—by the psychiatrist Tomio Hirai (1975). *Concentration* could be defined as *enhanced attention, in which almost all contextual awareness is lost, and ideas, images, and facts relevant to the target are easily and even creatively called to mind, all accompanied by a sense of deep satisfaction and an ability to easily dismiss fatigue.* The mind is temporarily dedicated to the target. This is the style of thinking, perceiving, and functioning referred to as *flow* (Csikszentmihalyi, 1990), something like what happens when people get deeply immersed in a hobby or when they are pursuing a particularly prized interest, even late at night.

THE FUNCTION OF ATTENTION AND INATTENTION—AN OPS PERSPECTIVE

Good attention is absolutely fundamental to good thinking; attention is a *precondition* of thinking (Lezak, Howieson, & Loring, 2004). Therefore, difficulties with the quality of attention can have serious, limiting repercussions all down the line for individuals, organizations, and society. That can affect prosperity and ultimately education funding in a downward spiral (such as

PISA testing, perhaps, and trying to compete with Asian automobiles like Lexus, which have risen to the first rank). *Attention deficit* could be defined in the ops approach as *a capacity for sustaining or shifting focus that too often falls below the threshold required to carry out necessary tasks and to fulfill responsibilities, making effective operating difficult.*

That term *too often* means that the attention deficit isn't *always* present. Many people who have difficulties with attention control find that the difficulties occur on a dimension of intrinsic interest. Activities that rank higher on the intrinsically interesting dimension may be associated with a negligible attention deficit—the target is fascinating to the person all by itself, so attention control is not an issue. For example, watching an interesting movie or a sports program might be associated with almost zero attentional deviation from the target. Activities that rank lower on the intrinsically interesting dimension such as, for most people, math or logic exercises, or foreign language memorization, may be associated with a marked difficulty controlling attention.

That may be why traditional Asian training in attention control typically chooses as a target something neutral (i.e., boring), such as a rock in a harmoniously arranged, raked rock garden, or the word *om* repeated over and over in an echoing room, or counting (Hirai, 1975; Paul, 2000). The quality of attention control cannot be adequately assessed or improved if the target is highly attractive to begin with. It is in attempting to calmly attend to neutral stimuli over a period of years (e.g., in an ashram, dojo, or even an elementary school, learning grammar and arithmetic) that the quality of attention control can be assessed and developed.

And that may be how the laissez-faire approach to child-rearing and education (e.g., Dewey, 1998; Neill, 1972; Rousseau, 1762/1993; Spock, 1957), conceivably, could contribute to a societal decrement in attention control. Children are naturally restless and curious as well as easily bored and distracted. This set of traits is plausibly selected for in survival—thus, there is a certain survival benefit to the inattention tendency. This is the positive, flip-side interpretation of attention-deficit disorder—an *active-curiosity tendency*, it might be called—a proclivity that is actually constructive (Piaget & Inhelder, 1969). Being on the move, checking on what is around, and persistently exploring new things apart from announced objectives seem valuable in helping to develop children's motor and visual abilities, building their knowledge base, helping to discover food sources, as well as helping to prevent them from being ignored, left behind, or unexpectedly attacked. But a critical skill set for adults is largely the reverse of this roving activity and impulsively shifting attention.

Adults, to be successful, largely need to be capable of focusing on one task aimed at one objective, then to follow through, one task after another in a sequence, all the way to completion with reasonable efficiency despite distractions and fatigue, within a larger series of projects aimed at survival. The timeless examples of farming, building houses, caring for animals, collecting wood and food, cooking and tending a fire, and cleaning, not to speak of attending to children, all call for robust attention control. More modern examples of situations calling for attention control would be learning reading, linguistics, mathematics, history, or foreign languages, managing a store or restaurant, preparing a law case, doing surgery, or flying an airplane.

All that implies that there needs to be a *transition* from *restless exploration* to *comfortable concentration* if the maturation process is not to derail into AD/HD. Human nature being what it is, that means children frequently are going to *resist* making the transition from a certain intuitively resonating *license* to what perhaps can be thought of as an *acquired taste* for attention control. This is something like the shift from primary learning, which is thrilling and compelling (like walking), to secondary learning (like math), which, at first, is not (Geary, 1995). That is where the parents and teachers come in—attention control is to an extent a *training* issue. And that is part of the rationale of the operations approach, to clarify the idea that an output, like focused attention, is a result of *building a system*, component by component, layer by layer, that will reliably generate that output. The operations premise is that if 50 percent of the mind has to be taught (Polderman et al., 2015), then that needs to be done by, initially, the environment—such as parents, teachers, and coaches. It is probably unrealistic to assume that children, on average, will master difficult operations on their own initiative—a beguiling but unlikely occurrence that might be termed the *Rousseau illusion*.

TYPICAL CLINICAL COMPLAINTS REGARDING ATTENTION CONTROL

Children and the environment (e.g., parents, teachers, and coaches) thus often conflict around attention management. Regarding AD/HD, children and parents often have diametrically opposed complaints. The young people are on a roll, as far as they are concerned. The AD/HD problem from an adolescent point of view is that the parents and schools are expecting a degree of focus and follow-through that is uncomfortable and unrealistic. The AD/HD problem from a parental or school point of view is the contrary—that the children need to learn to focus comfortably on difficult material, such as algebra, and there is no easy way around that. This generational conflict is partly a difference of *attribution*—determining the source of an issue.

It is not an issue of intelligence per se, as materials such as *Smart but Scattered* (Dawson & Guare, 2009) point out. People can have, for example, large vocabularies or large stores of information (a strong IQ) and still function in an inattentive fashion. But there does seem to be a leitmotif of *accomplishment* complaints, aside from IQ, floating through discussions of AD/HD by parents, children, and schools—an underlying *operational* concern.

As is sometimes ironically said, "Home is the place where they have to let you in at night." In that vein, schools are legally required to maintain a relatively *in loco parentis* role in relation to young people until adulthood is achieved at eighteen years of age. Therefore, although any resistance children may display to acquiring attention and production skills (i.e., operational competence) may be addressed by the schools, that resistance nonetheless is ultimately *protected,* and thus in a way *disguised.* Because failure has minimal real consequences until eighteen years of age, young people can resist attention control expectations with minimal consciousness of the long-term risks they are running.

However, that situation radically changes when children reach eighteen years of age. At that point parents and the school system tend to back away from direct control and development of character, attention, and other mental qualities, and their legal responsibilities usually end. Responsibility for self-management transitions to the young people *themselves.* Now, unexpectedly, they are in a radically different situation. Now, if their performance within a system (like the military, a job, or college) repeatedly falls below a standard due to AD/HD, they can be *ejected* from the system. This can have consequences that seem to be astonishing to almost everyone, including the public, such as poverty, or homelessness, or having to move back in with parents (i.e., boomerang). There is often a huge loss of young people's self-esteem and self-confidence. Jail is sometimes a consequence. Resumes are often spoiled, or at least clouded.

Adults with AD/HD, even young adults, eventually tend to transition over to what was previously just the *environment's* view of AD/HD, the attribution of parents and teachers. Adults with AD/HD eventually can see all too clearly the difficulties associated with poor attention control, insufficient follow-through, and general disorganization (Kolberg & Nadeau, 2002). As an adult, it is quite clear that attention is directly linked to successfully completing tasks and receiving financial compensation. They usually want to understand the condition, and they want to get better; they want to build a mental system that will control attention.

PORTRAIT OF HIGH-FUNCTIONING OPS PERSONALITY IN RELATION TO ATTENTION CONTROL

In all the lectures M gave to James Bond, M never said anything like this, "Now, see here Bond, you've got to pay attention. Moneypenny says you can never keep your assignments or your plane tickets straight. And Q says you never use any of his gadgets because you lose them. And Felix Leiter says you are always late to your appointments with the CIA. Pull yourself together, 007!" That conversation never occurs because it never could occur—a secret agent with AD/HD is unimaginable; someone like that wouldn't last. The fictional James Bond is a lot like his author, the sophisticated, daring, high-level Naval intelligence operative Ian Fleming (Lycett, 1995); such a mentality is the reverse of AD/HD.

High-ops people are calm yet attentive. They can focus on even a boring stimulus for hours with little difficulty. And they can easily shift attention from one target to another and back again as they choose. They understand the critical importance of controlling attention and find the process challenging and satisfying. They are well-oriented and well organized, and always have necessary materials at hand. They know what objective they are trying to achieve, the steps required to achieve it, and the larger purpose it serves. They know who they are, what they are doing, when they are doing it, and they do it well.

RECALIBRATION OF OPS COMPONENTS: ATTENTION CONTROL MAKEOVER

Medications quite often are helpful in managing AD/HD thought processes and behaviors, to the point where a medication evaluation might be the first consideration in dealing with an AD/HD case (Ivanov & Newcorn, 2005). However, it is also important to focus on *building strong psychosocial processes in clients in order to achieve long-term functional improvements* (Connors, 2006). That is the purpose of the ops approach.

- *Ideals:* The life, and the brain, have to be *managed.* That management is an *executive function* (Barkley, 2005). Deficiencies in attention control are often attributed to deficiencies in executive functions. "The executive functions consist of those capacities that enable a person to engage successfully in independent, purposeful, self-serving behavior" (Lezak, Howieson, & Loring, 2004, p. 35). In other words, successfully managing the mind, including attention, is fundamental to success.

  There is a certain reassuring, charming appeal to people who are overtly directionless and marginally effective. Woody Allen has made a lucrative career out of portraying lovable but hapless characters of dubious executive competence (Lax, 1992). His idol, the Swedish filmmaker Ingmar Bergman, ironically has made gorgeous, high-end films about similar people—directionless, restless, rootless, vaguely depressed, self-defeating, and lonely (Bergman, 1972). But are these good *role models*? Is it good to *idealize* these sorts of people?

Probably not; in reality, apart from their art, Allen and Bergman, as writers, directors, and producers, are probably very good at logistics; seem to be energetic, ambitious, and organized; appear to know exactly what they are doing; and are quite successful. So, an initial step in preventing, or dealing with, AD/HD is to consciously admire competence in attention control. That means *admiring calmness, focus, follow-through, efficiency, and resistance to distraction* as personality qualities. That might seem a little bland, a little boring, a little déclassé and un-chic in relation to the current zeitgeist of colorful mayhem. Nonetheless, the first step in the ops approach, as usual, is a conscious *value shift in a preferred direction* strong enough to resist unhealthy trends (i.e., what might be considered *ideational toxins*) in the culture.

- *Cognition:* The operations approach to living and thinking is aimed at strengthening cognition and reducing mental difficulties, such as AD/HD. As with so many issues regarding the mind, surface structure can be almost infinitely complex and therefore marvelously confusing, while the deep structure can be shockingly simple. If the basic process of living is setting goals and pursuing them—carrying out operations—while paying attention to them, then that needs to be a core of development. It is not really that hard for parents to remind children, or clients to remind themselves in self-talk, to focus on that basic two-step, means-to-ends plan: (1) "What do you/I need or want?" and (2) "What do you/I need to do to get there"?
- *Feelings:* Planning (such as getting homework done), and *subordinating to the plan momentary alternative attractions* (such as watching television), is something of a secondary (difficult) skill (Geary, 1995). Because often it is not automatically acquired, it may need to be taught. If the child, and later the adult, is to acquire planful, and capable, *self-efficacious* behavior (Bandura, 1982) outside of direct parental supervision, then the parent needs to build in to the child's thinking *intrinsic motivation for operational competence*. That can be done in a patient, simple way, with *positive prompts and redirects* to get the planful activity started (e.g., "You know what might be fun?)," and then *positive reinforcements* when tasks are competently completed (e.g., "Good job!"). Clients can acquire the habit of doing the same for themselves using self-talk. Focus and follow-through thereby become enjoyable and *self-reinforcing*.

Immersion in relatively fun tasks that nonetheless demand close attention (as opposed to passively watching electronic screens) can also help in attention training. For example, Sherlock Holmes mystery stories call for close attention to clues and reasoning. Crafts and hobbies, such as building workable train or plane models, or paint-by-the-numbers kits indirectly train attention, as do complex games such as chess and go, and crossword puzzles.

- *Decisions:* A lot of habits, once they get going, take on a life of their own, so to speak. Giving in to distractions is a little bit like that. If it gets well established past, say, the age of six years, it can be very difficult to haul the train back onto the tracks. That is because the brain develops in various ways in order to do the jobs that it is given (Nelson, de Haan, & Thomas, 2006). If the brain is not asked to pay attention, it may begin to lose that ability. Thinking and the personality crystalize year by year until a particular personality and thinking style is more or less established around age twenty years (Kaufman & Lichtenberger, 2006; Kernberg, Weiner, & Bardenstein, 2000). So, an AD/HD diagnosis means that thinking and attention have gotten a little off-track. It helps if that is *recognized* by the people involved, and then a conscious *decision* is made to get thinking, attention, and functioning consistently back on track. The sooner the better.
- *Actions:* There are numerous guides for doing that, such as the very accessible *ADD-Friendly Ways to Organize Your Life* (Kolberg & Nadeau, 2002), and the systematic, precise, thorough *Retraining Cognition: Techniques and Applications* (Parente & Herrmann, 1996).

  For a family and for an individual with AD/HD, all this means taking activities a little slow; stopping to think about objectives and tasks; planning in a very straightforward, simplified, concrete way; putting everything on a schedule; posting the schedule; and following the schedule until tasks are completed. There needs to be a conscious effort to keep everything super-organized, with lots of notes and reminders. Focusing itself needs to be a focus of attention; it needs to be consciously established and maintained as the zeroth priority—*remember to remember* might be the mnemonic. And then, lots of encouragement and praise as activity and attention are maintained on a narrow track, because staying focused on one line of activity is difficult for AD/HD people.
- *Objectives:* The trick, again, is to keep it simple, concrete, reachable. AD/HD clients are forever wanting to climb Mount Kilimanjaro at a single bound; then Everest; then Kilimanjaro again. That doesn't work, so that is discouraging, and adds to the chaos. Manageable objectives work better.
- *Communication: The New Yorker* magazine is a famed bastion of good writing, amusement, and general sophistication. It is particularly noted for long, discursive pieces that meander hither and yon in an entertaining way. Just so, an interviewer asked the *New Yorker* editor, David Remnick, a simple question: "Unfathomable . . . the roughly seven-minute answer is as entertaining, erudite and occasionally baffling as the magazine itself" (Martinson, 2016). AD/HD communication can be a little like that.

One therapeutic idea for AD/HD people would be to learn to communicate a little more like General Patton! His book is, in its own way, a fairly good model of focused communication—concise, factual, and purposeful, as well as being fascinating and ultimately, perhaps, motivating (Patton, 1947). Exercising the *communication discipline* of habitually *using language closely relevant to a purpose* can help prevent the derailing of thought and action from a selected direction.

SUMMARY

The culture itself may have veered off a focused and productive track. And that cultural shift may have influenced parenting and self-management in the direction of self-indulgence and impulsivity. Ergo, there seems to be something of an attention-deficit epidemic in society. The process, then, for those who wish to function in a more focused manner, to become more operationally astute, is to consciously prize and then practice following a critical path to chosen objectives despite distractions until excellent concentration becomes almost second nature.

FURTHER CLINICAL CONSIDERATIONS

There is a surprising comorbidity of AD/HD and addiction. That may be partly because, at a level up in abstraction, they are similar—getting off-track, spending time, effort, and attention on interesting activities not closely linked to important goals, which then often end up contributing to self-defeat—*shattered operations*. Improving attention control assists addiction recovery and the reclaiming of a sane and satisfying life.

Chapter 18

Addictions

Toxic Rose-Colored Glasses

Farewell, My Lovely is a classic noir novel featuring the private investigator Philip Marlowe (Chandler, 1940/1992). The excitement begins with a mysterious murder. In the process of investigating that, Marlowe visits the wealthy and beautiful Mrs. Grayle at her huge home and asks her some questions. In a long, rolling exchange of racy repartee, Marlowe and Mrs. Grayle appear to consume about eight glasses of scotch each. If a shot of whiskey is 1.5 ounces, and if a fifth of scotch is about 25 ounces, then in the space of perhaps an hour Marlowe and Mrs. Grayle went through a bottle of scotch. That is a lot of alcohol. Not surprisingly, Raymond Chandler was an alcoholic (Day, 2014). But he made that amount of drinking not only *seem* completely normal, but fashionable. Of course, for Chandler, it *was* normal. And for Marlowe, it *was* fashionable.

Interestingly, Marlowe lived the kind of life that clinically often appears associated with excessive alcohol use. He lived alone in a small, inexpensive apartment. He ate only enough to get by. He never talked about his parents or home—he seemed to have no roots. He did not have a girlfriend. He did not seem to have any friends or hobbies. With all due respect to the noir novel genre, he did not quite have a job—just dangerous and confusing pick-up work now and again for which he often was not paid. He was relatively poor. His perception of Los Angeles was a little austere and ungratifying; it seemed to consist of an endless, sun-bleached highway rolling through faded houses and dangerous, broken-down neighborhoods populated by strange people of uncertain intent. Naturally, all that makes for great *reading*. Nonetheless, the life described likely would be stressful. And *addictions are all about avoiding stress.*

DEFINITION OF ADDICTION

From an ops perspective, *addiction* could be defined as *compulsive reliance on a substance or activity to avoid experiencing painful emotions.* The stress-relieving substance or activity is a little like the sorcerer's apprentice—it starts out helpful, and then it takes over the addict's life. Not always, of course. But usually there is a steady rise in the amount of consumption as tolerance rises, the addiction becomes progressively more important, the costs go up while the price the addict is willing to pay rises higher and higher, and the capacity to face the situation and deal with it in a straightforward manner falls lower and lower until, often, life goes over the edge at a catastrophe point (Zeeman, 1976). The addict can lose friends, spouse, children, car, house, job, savings, health, and even freedom, not to mention self-respect (e.g., the evocative stories of these tragedies in *Alcoholics Anonymous*, 1976).

There are several quick alcohol screening tests, like the CAGE (an acronym of key issues: Cut down-Annoyed-Guilty-Eye opener) and the AUDIT (Alcohol Use Disorders Identification Test) (National Institute on Alcohol, n.d.). And there are more thorough evaluations of the personality itself, like the MMPI (Minnesota Multiphasic Personality Inventory) that keys in on addiction as well as the personality components that might be contributing to an addiction. The limitation of a short list of screening questions like the CAGE is it misses a lot. But the strength of some of these quick assessments is that they consist of several low-probability questions. That is, almost *no one* without an addiction would answer *yes* to a question, such as, "Eye opener: Have you ever had a drink first thing in the morning to steady your nerves or to get rid of a hangover?," but people with an alcohol addiction often answer *yes* to that question. The probability that someone would answer *yes* to *several* low-probability questions in the absence of an addiction is low indeed. Therefore, these quick screening devices can be useful in determining whether a line has been crossed from alcohol use and even from very high use (abuse) to addiction. This is helpful for family, for friends (as with interventions), and even for addicts themselves.

But some people argue that this is all neither here nor there. People indulge in many substances and activities that lend themselves to excess, such as drinks after work (alcohol), calming medications like benzodiazepines (drugs), fun weekends in Las Vegas (gambling), strolling through malls, looking for the best sales (shopping), trying out the latest restaurant (food), or closely following news and sports (computers). Rather than trying to draw a fine and persnickety line between social use and excess, then moving the line over if it looks like it is about to be crossed, a more fundamental question to ask is: "*Is this substance or activity resulting in serious difficulties in my life?*" If the answer is *yes*, then it might be time for a change (Donovan, 1999).

THE FUNCTION OF ADDICTION: AN OPS PERSPECTIVE

Addiction might seem like an unalloyed flaw in the design for living, but it may have its virtues. The marvelous concept of *flow* (Csikszentmihalyi, 1990) says that there is almost nothing better in life than getting really *immersed* in some interesting and important activity, in a process of *deep concentration,* much like Zazen (Hirai, 1975)—similar to the addictive process. The concept of *expertise* (Puppe, 1993) says that really excellent output comes from a *long period* of narrow focus, developing knowledge and refining skills until they can be executed with almost effortless, on-target bravura—similar to the addictive process. Accumulating millions of dollars (Lynch, 1997) requires accepting the necessity of highly focused financial effort consistently exercised—similar to the addictive process. And the esteemed psychologist Dean Simonton (1994) proposed that *greatness* (George Washington crossing the Delaware was on the cover of his book) follows about *ten years of total immersion* in a skill and knowledge domain—similar to the addictive process. In other words, one aspect of addiction is a dedicated, narrow, lengthy, almost obsessive focus, a tendency that can be quite beneficial. (For clients, this amounts to something of a *reframe*.)

And, clinically, it is surprising how many people with addiction issues do radiate an aura of greatness—casual self-confidence, smooth social skills, savvy perceptiveness , talent, and a *je ne sais quoi élan* (an inexpressible lightness of being). The glitch in addicted people's systems is that their lives often are skewed toward evasion rather than accomplishment. They usually have a hidden fear, as suggested in these lines from the song "Darkness on the Edge of Town": "Everybody's got a secret, Sonny/Something that they just can't face/ Some folks spend their whole lives trying to keep it/They carry it with them every step that they take" (Springsteen, 1978). They often develop a hidden identity and a hidden life that revolves around protecting their addiction. That sets up parallel identities, sometimes termed *schemas* (van Genderen, Rijkeboer, & Arntz, 2012). Addicts become adept at shifting from one schema to another, trying to keep reality at bay, trying to keep from spoiling their public identity (Goffman, 1986). But all that skilled dramaturgy siphons off energy and talent as well as confidence and hope because the strategy is chronically risky. Usually, the truth comes out, and then, all too often, the secret schema consumes the public life (e.g., the Big Book, *Alcoholics Anonymous*, 1976).

TYPICAL CLINICAL COMPLAINTS REGARDING ADDICTION

Clinically, complaints in relation to addiction parallel the situation in AD/HD, two disorders that are frequently comorbid. People in the throes of addiction, but who haven't yet *hit the wall* (their limit as to how much difficulty due to addiction they are willing to endure), typically complain about the unfairness or irrationality of drug laws, or the bias of the police, or the insensitivity and unreasonableness of spouses, supervisors, and judges. This appears to indicate, again, an attribution paradigm, in which the source of the issue seems to be in *other* people, not in the addict.

Contrariwise, the environment (friends, family, therapists, any involved officials, the workplace) is looking around, and in the scores of people nearby, there appears to be only one—the addict—who is routinely having difficulty controlling alcohol, food, shopping, gambling, or whatever the addiction is. So, the environment attributes the difficulty to the addict, and it is the addict who needs to change, not the environment.

Most environments, unlike the wonderful organization Alcoholics Anonymous (AA) (*Alcoholics Anonymous*, 1976; *Twelve Steps and Twelve Traditions*, 1953), do not usually demand complete abstinence. Most environments simply ask the addicted person to back off some, enough to function adequately in a basic way—go to work every day, pay the bills, and take the family on more picnics would be about the size of the environment's wish list. The reason is something that might be called the *duality of addiction*: (1) narrow dedication to one line of activity does have its benefits (such as mastering STEM fields of knowledge—science, technology, engineering, and mathematics), and even a *craving* for such activities is OK; however, (2) it is not really wise to carry anything, even study, hard work, exercise, or the accumulation of wealth to the point of sizable costs in health, career, or family life. Many aspects of life that are interesting, exciting, satisfying, or potentially lucrative can be carried to an extreme. It is not so much the aspect of life that is the difficulty, as the *taking of that aspect to a self-destructive extreme*. Thus, developing a personality well able to manage threat and fear, responsibilities and stresses, desire and rewards is the underlying issue.

PORTRAIT OF HIGH-OPS PERSONALITY IN RELATION TO ADDICTION

High-ops-type people like to have fun in life (e.g., Beavers, 1977; Bonington, 1981; Epicurus, 2012; Loren, 2014). They like to work hard and play hard. They pursue deep involvement that results in high competence and substantial sophistication (Simonton, 1994). But they also know that immediate activities always fit within a larger, *primary frame of life*. Their lives are framed by *desiderata of ultimate importance*—health, family, home, education, career, friendships, reputation, and financial viability, among others, those essentials that make life work.

High-functioning people carefully monitor their thoughts and behavior around a risk-reward correlation; they are just not attracted to high but pointless risk. For example, gambling can be fun. People frequently go to Las Vegas and earmark a certain amount of money for gambling, say, $100. They are willing to spend that much just for the fun, part of a silly weekend in Vegas—a *phenomenological* payoff. But they know perfectly well that the probable *financial* payoff in gambling is zero. Therefore, it wouldn't be rational to bet the farm on a spin of the roulette wheel. So, they don't. That keeps them on the high side of the risk-reward correlation—they took a small financial risk and received in exchange a moderately high phenomenological payoff. They are able to have fun dining, dancing, and strolling through the streets and seeing the sights without getting *hooked* on anything, and are able to return to their larger frame of life Monday morning in reasonably good shape. Altogether, they lead exciting, interesting lives, but *remain the masters of them.*

RECALIBRATION OF OPS COMPONENTS: ADDICTIONS MAKEOVER

- *Ideals:* Probably, the essential protection against addiction is having loving parents (Beavers, 1977). Parents can make many mistakes, but if they genuinely love their children, and if the family has fun together, then the children will tend to forgive those mistakes. But what to do if, as in the typical case, the parents themselves had difficulty in genuinely caring for a child, a child who then develops an addiction, as in a family system of multi-generational difficulties (Bowen, 1978)?

  Research on resilience (Wolin & Wolin, 1993) suggests there is a way, not really all that different from AA and twelve steps. The idea is to not give into the lure of cynicism, but to *maintain a belief and trust in life and people* (not too much, of course—naiveté is the flip side of cynicism). Resilient people consciously strive to maintain their faith in life and themselves as they adopt a pragmatic approach to living well. In particular, they actively seek out people they can respect, believe in, and trust for friendship and guidance—they develop a *surrogate family*. The Big Book (*Alcoholics Anonymous*, 1976) is as much general spiritual guidance as it is practical suggestions for managing an addiction.
- *Cognition:* Human beings are surprisingly hardy. As the anthropologist Robert Ardrey (1967) put it, people are the *bad-weather animal*. In other words, people can cope remarkably well with a great deal of adversity. Unfortunately, therefore, they can do a lot of damage to themselves and survive. Young people usually don't know anyone who has died. But they

often know a lot of people who have taken serious risks with drugs and alcohol, driving and even violence, and come through it all with about as many obvious long-term repercussions as Captain America. Perhaps few attorneys outside the Supreme Court are as skillful in defending addictions as the average addict not yet in recovery. Those clever defenses tend to prolong addiction and increase the risks.

However, ultimately addiction doesn't seem to be a cognitive issue. There can be numerous close brushes with disaster, good friends can fall by the wayside, doctors and judges can make dire warnings, and costs can mount up catastrophically—loss of family, job, and freedom. In other words, eventually, it becomes *obvious* that the addiction isn't a good idea. But, up to the point of recovery, those facts and logic do not matter. In the dance of addictions, emotions call the tune.

- *Feelings:* Mark Twain wrote a wonderful book titled *The Prince and the Pauper* (2006). A young prince and a remarkably similar-looking pauper of the same age happen to meet, exchange clothes, and then accidentally get separated. The prince wanders off to the pauper's poor home. And the pauper, amid much tension and second-guessing, tries to play the role of prince without getting caught. That is how addicts feel—like imposters. Addicts often heard negative messages from their parents and absorbed those; those formed the unwanted schema of their identities and their lives. Addiction helps maintain a distance from those early negative self-definitions. As long as they are participating in the addictive process, they feel good, like a prince or princess. But no matter what they do, they can never quite escape those underlying, negative, pauperized feelings, which forever hover at the edges of consciousness like malevolent wraiths.

  That is why recovery programs and AA go out of their way to help people feel wanted, to create a *cushion of belonging*, so that clients can begin, first, to face themselves honestly, and, second, to substitute positive values, beliefs, habits, and relationships for negative ones, and, then, finally, to substitute success and genuine self-esteem for shame and self-defeat.
- *Decisions:* Decision-making is a huge issue in addictions. It might almost be said that people who develop addictions don't *make* decisions; rather, decisions are made *for* them, first by the substance or activity and, then, eventually, especially after a catastrophe point, by the environment, such as a spouse, employer, or judge.

  The reason for that may be developmental. Children are naturally curious, reactive, and a little transient, rather than consistently being thoughtful, decisive, and careful to follow through all the way to a completed objective. To put that observation one way, the development of *will—follow through despite distractions*—is partly natural and partly has to be taught by parents and schools (Assagioli, 1973).

In addiction-prone families that is exactly what does *not* tend to happen, first, because the parents are preoccupied by their own stresses (Bowen, 1978) and, second, because (as most dieters know) development of the will is something of a secondary learning process (Geary, 1995). As such, development of the will is often uncomfortable for the children; therefore, they resist the parents. For the parents to be effective in teaching decision-making and will power despite resistance, the parents themselves have to be adequately concerned about the issue and also have to have will power of their own! If parents are struggling with issues of their own, such as addictions, the necessary teaching of the children may not occur.

Indeed, addicts often feel, underneath it all, like adolescents of advanced age who have trouble making good decisions and then following through adequately, leading to their characteristic underlying sense of helplessness and shame. Clinically, a large portion of recovery programming, often covert, is about learning to make better choices and abide by them, and thus developing confidence and mature self-respect (e.g., *Twelve Steps and Twelve Traditions*, 1953). The twelve steps themselves are essentially about making the right choices.

- *Action:* Addicts are sometimes scandalized by and amused by the disarray of their lives (e.g., Burroughs, 1977). The addiction process is in many ways a *deviation* from sensible self-management. The process is all about feeling good at almost any cost, and that usually does not include neatening up, careful labeling and filing, cleaning, budgeting and record-keeping, laundering, scrupulous grooming, or any of those other proper organizational habits espoused in such works as *How to Be Organized in Spite of Yourself* (Schlenger & Roesch, 1989). Another subtext in good recovery programs, despite the frank, family therapy atmosphere, something like Virginia Satir's systems communication training (1967), is often a Schlenger and Roesch-style teaching program of good grooming, money management, and personal organization. Addicts putting their lives back together is not unlike the process described in a basic job analysis text (McCormick, 1979)—analyze the job (of living), establish the job requirements, design the best way to fulfill those requirements, follow the design. People in good recovery are usually pleased with their capacity to manage their lives in a sensible, step-by-step fashion (similar to the twelve steps) aimed at maintaining sobriety. Sobriety and the sensible activities it involves becomes a job they are proud of.
- *Objectives:* Similar to the outlook and approach of people diagnosed with AD/HD, those with addictions often find their reach vastly and depressingly exceeding their grasp. Their view is not too far from Oscar Wilde's (1986, p. 121) ironic suggestion that, "Moderation is a fatal thing . . . . Nothing succeeds like excess." Reining in that over-ambition is often a

painful part of addiction treatment, as it seems to entail the loss of all those prized longings. But, in reality, those ambitions are often compensatory and illusory (Adler, 1973). A much firmer basis for success in life and for genuine self-esteem is to choose a small number of feasible goals, address those in a direct, step-by-step fashion, succeed one achievement at a time, and base plans and expectations on that record of accomplishments—a modest and doable approach to operations. Thinkers from Aristotle (2001) to Confucius (n.d.) to the Wolins (1993) have urged a sensible, *middle way* in life. As with Zen and other spiritual disciplines, a simple, direct, pragmatic path is often the best. *Alcoholics Anonymous* (1976) does a great job of clarifying the value of limited objectives and a simple life.

- *Communication:* Addicts are famous for having glittering banter and quick, plausible answers for almost any question (i.e., entertaining feedforward). The Welsh lyric poet Dylan Thomas, for example, was both an alcoholic and a renowned and cherished raconteur (Sinclair, 1999). That clever banter and that whole atmosphere of sophisticated living is itself intoxicating and functions as a defense against the sense of poverty underneath.

  One of the primary tasks of many recovery programs is to build a foundation of belonging and security so that addicts can drop the defenses that are blocking the painful but necessary reality information (i.e., feedback) that would encourage them to get back on track (Vaillant, 1977). *Denial* says that there is no problem whatsoever. *Minimization* says that there is a problem, but it is negligible. *Projection* says there is a problem, but it is someone else's. *Reaction formation* says that far from a problem, everything is great. *Humor* says, sure, there are problems, but the only thing to do is laugh; "Listen to this: two guys walk into a bar . . . ." Skilled addicts (which is to say, most) can flip around these defenses like nunchucks in a Bruce Lee martial arts movie, driving away all but the most dedicated helpers. That is why learning to *not* depend on defenses is so important. *Directness and candor are touchstones of recovery,* in programs, personnel, and materials. But for addicts those qualities represent very, very difficult challenges.

Taking off those toxic rose-colored glasses calls for a whole new way of seeing, living, thinking, and communicating. A new life.

SUMMARY

An addict might be thought of as a hero-manqué. Many of the qualities that go into greatness—deep immersion in a process, nervy willingness to take risks, utter determination to keep going despite the costs, an attraction to glitz and glamour, and a sizzling, convincing patter—characterize many addicts. The problem is that a lot of that street brilliance is dedicated to evading a profound sense of loneliness and loss. Serious addicts begin to live again when they leave all that behind, and begin to relate to themselves, to others, and to the world with simplicity and directness, responsibility and consistent effort, despite the pain.

Afterword

A great many books have been written on psychotherapy and personal change. My favorites are always those that have titles that include terms like "key," "secret," "trick," "magic," or "three steps to" or "five days to." I love the idea that in exchange for reading a couple of hundred pages, immense wisdom, unshakable serenity, youth and slimness, and perhaps even wealth and fame could be mine. I have a lot of those books.

What might be called the *philosophy of operations*, as presented in this book, is a little different. The operations view says that life is a competitive situation, and surviving and thriving calls for the persistent development and application of useful skills and knowledge—*the mastery of critical operations*. Those operations are in the realms of emotion and intellect, behavior and society, vocation and finance. That development is a life-long process. It begins with caring and conscientious, but firm and savvy parents, and is continued by the individual, and perhaps eventually becomes embedded in the culture on behalf of future generations. Because the development process is frequently demanding, it needs to be conducted with a positive attitude, with a certain creative, affectionate, game-like bravura if life is to be a joy rather than a burden.

If critical operations are not mastered, that can result in a panoply of miseries large and small, almost none of them easy to correct (if they were easy to correct, as the psychiatrist Jerome Frank pointed out, they would have been corrected long ago). If serious discomfort is experienced in living, that implies a reconstructive process, a recalibration of the personality. That calls for courage and commitment, patience and methodicalness, a willingness to get lost in the psychic forest, in the dark, and to keep going, to reach that shining city on the hill—one's best self.

The ops approach to understanding the self and to living successfully is not *Mission Impossible*, but it is not *Mission Easy*, either. Nonetheless, the signature invitation of *Mission Impossible* may be appropriate to the quest that is mind ops—the opportunity is open to anyone, it is only a question as to whether "you choose to accept it."

Appendix A

Self-Assessment

1. *Ideals and Values*
   a. Have I identified high, overall principles for living? Can I say what they are?
   b. Are my basic principles just and compassionate so that all benefit?
   c. Do my basic principles work for me, too? Am I taking good care of myself?
   d. Do I have ideals that are a bit of a reach, that challenge and inspire me?
   e. Do I have reasonable day-to-day values that help me feel good about my life and that I live up to almost without fail?
2. *Cognition*
   a. Do I think in terms of verifiable facts?
   b. Do I link those facts logically, so my conclusions are reasonable?
   c. Do the conclusions I arrive at work pragmatically, and are they also, where possible, respected by others?
   d. Do I consciously strive to make my thinking precise, numerical, sequential, and goal oriented?
   e. Am I developing repertoires of skill and knowledge that allow me to succeed more and more in life at the goals I choose?
3. *Feelings*
   a. Do I know what I feel? Am I in touch with my feelings?
   b. Do my feelings show a fairly clear link with reality, so that if things go well I feel good, and if things don't go so well I feel less good?
   c. Can I tell other people what I feel candidly if they have a need to know and if I want to tell them?
   d. Do I make a point of designing and experiencing a stimulating and satisfying sensory life—colorful and richly textured, beautiful and serene?

   e. Do I focus on positive emotions, and savor my successes, yet face forthrightly negative emotional situations and deal with them as smoothly as I can?
4. *Decisions*
   a. Do I make a point of thinking in terms of means and ends?
   b. Do I primarily base decisions on facts and probabilities, while taking into account feelings?
   c. Do I routinely and honestly consider costs as well as benefits?
   d. Do I consider possible long-term consequences as well as immediate outcomes?
   e. Am I willing to reflect without getting stuck, and select without being impulsive?
5. *Action*
   a. Do I consciously link action to savvy evaluation and careful decision-making?
   b. Do I make sure my actions are oriented toward goals? (Unless I'm just having fun.)
   c. Do I usually sequence my actions rationally—first things first?
   d. Do I start promptly and follow through to completion despite difficulties?
   e. Do I hone within myself a confident, game-like attitude that allows me to strive mightily toward success while taking failure in stride?
6. *Objectives*
   a. Am I aware of my needs, my wishes, and my dreams, and do I translate those into specific objectives?
   b. Do I set objectives within a reachable range—interesting and challenging, but not impossible?
   c. Do I identify natural milestones, so that I can monitor my progress toward my objectives and move toward my objectives in relatively easy stages?
   d. Do I establish start-finish deadlines so my objectives aren't just Someday Isles floating in a Fantasy Sea?
   e. Do I make a point to really celebrate successes with friends and family?
7. *Communication*
   a. Am I putting the best face on my ideas, plans, and projects?
   b. Am I also willing to listen to what others have to say about me and my activities with sincerity and minimal defensiveness?
   c. Do I speak and write accurately, concisely, and gracefully while being persuasive?
   d. Do I use communication for the benefit of others as well as myself?
   e. Do I relate to others with a certain *je ne sais quoi flair*—informed, light, warm, close, amused, and amusing while being serious and idealistic?

Appendix B
Operations Personality Types

1. *Ideals type*: Highly considerate. Always concerned about others and the right thing to do. Stands up for other people; wants to make sure all people get their needs met. Interested in philosophy, spirituality, and the kind of literature and discussions that address larger issues of belonging, compassion, and justice. Reliably kind.
2. *Cognition type*: A thinker. Always looking at the facts and logic of a case. A stickler for precision. Well-informed, with a hunger for good information. Not intimidated by complexity or mathematics. Dives into intellectual difficulties and straightens them out. Reliably accurate.
3. *Feelings type*: Excited about life. Loves color, romance, the scent of the sea, and the sound of thunder. Passionate about art, music, people, stories, travel. Sensitive and deeply emotional, thoughtful and expressive. Events in life register with a strong impact. A poet at heart. Relationships are the key issue.
4. *Decisions type*: Slightly removed from life, in an attitude of cool appraisal. Very alert, even-handed, and sensible. Considers situations and options from all relevant angles. Independent—cannot be pressured into choosing. But confident in making a choice and standing behind it. A leader.
5. *Action type*: Gung ho. Loves getting going. Not afraid to jump in and wrestle with issues, conflicts, and people. Good at building action alliances. Good at project organization—sequencing tasks, developing milestones, pushing through to the end. An adventurous spirit. A good buddy, employee, team member. Life is a game, so *let the games begin!*
6. *Objectives type*: The dreamer that guides the doer. Life needs a mission to be coherent and satisfying—this is the person who prioritizes that. That sterling diploma, the great career, the beautiful family, that brilliant idea, financial independence, that house overlooking the ocean or the valley, the

great American novel. This is the person who takes seriously the possibilities of the future. The practical visionary. The Steven Jobs. The Susan Anthony. The Walt Disney.

7. *Communication type*: Charming and persuasive. Yet sincere and honest. Writes well and speaks well. Loves getting together with other people and discussing ideas, plans, projects, and activities. Fun and interesting, even funny, but also a good listener, rarely defensive—great company. The spokesperson. The diplomat.

References

Adler, Alfred. (1973). *Individual Psychology*. Totowa, NJ: Littlefield, Adams.

Aeschylus. (1961). *Prometheus Bound, the Suppliants, Seven Against Thebes and the Persians* (Philip Vellacott, Trans.). Harmondsworth, England: Penguin Books.

Albus, James. (1981). *Brains, Behavior, and Robotics*. Peterborough, NH: BYTE Books.

Alcoholics Anonymous. (1976). New York, NY: Alcoholics Anonymous World Services.

Aleksander, Igor. (2001). *How to Build a Mind: Toward Machines with Imagination (maps of the mind)*. New York: Columbia University Press.

Alford, Brad A., & Beck, Aaron T. (1997). *The Integrative Power of Cognitive Therapy*. New York, NY: Guilford Press.

Alger, Horatio, Jr. (2008). *In a New World: Among the Gold-fields of Australia*. Teddington, England: Echo Library.

American Psychiatric Association. (1968). *Diagnostic and Statistical Manual of Mental Disorders* (2nd ed.). Washington, DC: Author.

American Psychiatric Association. (2000). *Diagnostic and Statistical Manual of Mental Disorders* (4th ed., text rev.). Washington, DC: Author.

Apollonius. (1959). *The Voyage of the Argo* (E. V. Rieu, Trans.). Harmondsworth, England: Penguin Books.

Ardrey, Robert. (1967). *African Genesis: A Personal Investigation into the Animal Origins and Nature of Man*. New York, NY: Dell.

Aristotle. (1961). *Aristotle's Poetics* (S. H. Butcher, Trans.). New York, NY: Hill and Wang.

Aristotle. (2001). *The Basic Works of Aristotle* (Richard McKeon, Ed.). New York, NY: Modern Library.

Assagioli, Roberto. (1973). *The Act of Will*. Harmondsworth, England: Penguin Books.

Aurelius, Marcus. (2006). *Meditations* (Martin Hammond, Trans.). London, England: Penguin Books.

Auspitz, Oliver (Producer), & Dornhelm, Robert (Director). (2008). *La Boheme* [Motion Picture]. Germany: UNITEL.

Austen, Jane. (2010). *Pride and Prejudice* (Annotated ed., Patricia Meyer Spacks, Ed.). Cambridge, MA: Harvard University Press-Belknap. (Original work published in 1813.)

Axelrod, Alan. (2000). *Elizabeth I, CEO: Strategic Lessons from the Leader who Built an Empire.* New York, NY: Prentice Hall Press.

Ayer, A. J. (1952). *Language, Truth and Logic.* New York, NY: Dover. (Original work published in 1936.)

Aylwin, Susan. (1985). *Structure in Thought and Feeling.* London, England: Methuen.

Bacon, Francis. (1960). *The New Organon and Related Writings* (Fulton H. Anderson, Ed.). Indianapolis, IN: Bobbs-Merrill.

Bandler, Richard, & Grinder, John. (1975). *The Structure of Magic: A Book about Language & Therapy.* Palo Alto, CA: Science and Behavior Books.

Bandura, Albert. (1969). *Principles of Behavior Modification.* New York, NY: Holt, Rinehart & Winston.

Bandura, Albert. (1982). Self-efficacy mechanism in human agency. *American Psychologist, 37*(2), 122–47.

Bardot, Brigitte. (1996). *Initiales BB: Memoires.* Paris, France: Grasset.

Barkley, Russell. (2004). *ADHD and the Nature of Self-control.* New York, NY: Guilford.

Baron, Jonathan. (1985). *Rationality and Intelligence.* Cambridge, UK: Cambridge University Press.

Barrie, James M. (1995). *Peter Pan.* New York, NY: Barnes & Noble.

Barron, Fred (Producer), & Wolff, Art (Director). (1989). *Seinfeld, Season 1: The Very Beginning.* [Television Series]. Culver City, CA: Sony.

Basho, Matsuo. (1966). *The Narrow Road to the Deep North and Other Travel Sketches* (Nobuyuki Yuasa, Trans.). Harmondsworth, England: Penguin Books.

Baum, L. Frank. (1995.) *The Wizard of Oz.* New York, NY: Barnes & Noble Books.

Beavers, W. R. (1977). *Psychotherapy and Growth: A Family Systems Perspective.* New York, NY: Brunner/Mazel.

Beck, Aaron T., & Young, Jeffrey E. (1985). Depression. In David H. Barlow (Ed.), *Clinical Handbook of Psychological Disorders: A Step-by-step Treatment Manual* (pp. 206–44). New York, NY: Guilford Press.

Bergman, Ingmar. (1972). *Persona and Shame: The Screenplays of Ingmar Bergman* (Keith Bradfield, Trans.). New York, NY: Grossman.

Berne, Eric. (1961). *Transactional Analysis in Psychotherapy.* New York, NY: Grove Press.

Bertman, Stephen. (2010). *The Genesis of Science: The Story of Greek Imagination.* Amherst, New York: Prometheus Books.

Bevan, Timothy (Producer), Cavendish, Jonathan (Producer), & Maguire, Sharon (Director). (2001). *Bridget Jones's Diary* [Motion Picture]. Paris, France: Studio-Canal & London, England: Working Title.

Bird, Isabella L. (1998). *Six Months in the Sandwich Islands.* Honolulu, HI: Mutual. (Original work published in 1881.)

Blanning, Tim. (2011). *The Romantic Revolution: A History*. New York, NY: Modern Library.

Blinder, Martin (with Lynch, Carmen). (1989). *Choosing Lovers*. Macomb, IL: Glenbridge.

Block, Peter. (2000). *Flawless Consulting: A Guide to Getting Your Expertise Used* (2nd ed.). New York, NY: Jossey-Bass-Pfeiffer.

Blom, Philipp. (2011). *Wicked Company: Freethinkers and Friendship in Pre-revolutionary Paris*. London, England: Phoenix.

Bodenhamer, Gregory. (1983). *Back in Control: How to Get Your Children to Behave*. New York, NY: Prentice-Hall Press.

Bolles, Richard N. (1982). *What Color is Your Parachute 1982? A Practical Manual for Job Hunters and Career Changers*. Berkeley, CA: Ten Speed Press.

Bonaparte, Napoleon. (1992). *Napoleon on Napoleon: An Autobiography of the Emperor* (Somerset de Chair, Ed.). London, England: Brockhampton Press.

Bonington, Chris. (1981). *Quest for Adventure: 21 Stories of the Men and Women who Challenged Oceans, Mountains, Deserts, Snow, and Space*. London, England: Hodder & Stoughton.

Bossidy, Larry, & Charan, Ram (with Burck, Charles). (2002). *Execution: The Discipline of Getting Things Done*. New York, NY: Crown.

Bottero, Jean. (1992). *Mesopotamia: Writing, Reasoning, and the Gods* (Zainab Bahrani & Marc van de Mieroop, Trans.). Chicago, IL: University of Chicago Press.

Bourne, Lyle E., & Healy, Alice F. (2014). *Train Your Mind for Peak Performance: A Science-based Approach for Achieving Your Goals*. Washington, DC: American Psychological Association.

Bowen, Murray. (1978). *Family Therapy in Clinical Practice*. New York, NY: Jason Aronson.

Bowra, C. M. (1957). *The Greek Experience*. New York, NY: Mentor Book.

Brendan. (2010, September 24). Re: Anita Berber, Cabaret Berlin. [Blog post]. Retrieved from http://www.cabaret-berlin.com/?p=365

Brenner, Charles. (1973). *An Elementary Textbook of Psychoanalysis* (Rev. ed.). Garden City, NY: Doubleday-Anchor Books

British Broadcasting Corporation. (2015, October 14). "PISA tests: Top 40 for maths and reading." Retrieved from http://www.bbc.com/news/business-26249042

Brown, Donald E. (n.d.). Human universals, human nature, human culture. Retrieved from http://www.humiliationstudies.org/documents/BrownUniversalsDaedalus.pdf

Bruner, J. S., Goodnow, J. J., & Austin, G. A. (1956). *A Study of Thinking*. New York, NY: Wiley.

Buber, Martin. (1970). *I and Thou* (Walter Kaufmann, Trans.). New York, NY: Scribner's- Touchstone.

Burns, David. (1999). *Feeling Good: The New Mood Therapy*. New York, NY: Harper.

Burroughs, William S. (1977). *Junky*. New York, NY: Penguin Books.

Burroughs, William S., & Ginsberg, Allen. (1963). *Yage Letters*. San Francisco, CA: City Lights Books.

Calleros, Charles R. (2014). *Legal Method and Writing*. New York, NY: Wolters Kluwer.

Campbell, Joseph. (2004). *The Hero with a Thousand Faces*. Princeton, NJ: Princeton University Press-Bollingen. (Original work published in 1949.)

Camus, Albert. (1942). *The Stranger*. New York, NY: Random House-Vintage.

Caplan, Ruth (with Caplan, Gerald). (1969). *Psychiatry and the Community in Nineteenth-century America: The Recurring Concern with Environment in the Prevention and Treatment of Mental Disorder*. New York, NY: Basic Books.

Capra, Frank (Producer/Director). (1946). *It's a Wonderful Life* [Motion Picture]. USA: Liberty Films.

Carnegie, Dale. (1981). *How to Win Friends and Influence People*. New York, NY: Pocket Books.

Carroll, Lewis. (1946). *Alice's Adventures in Wonderland*. New York, NY: Random House.

Casti, John. (1989). *Alternate Realities: Mathematical Models of Nature and Man*. New York: Wiley-Interscience.

Cavanagh, Michael. (2006). Coaching from a systemic perspective: A complex adaptive conversation. In Dianne R. Strober & Anthony M. Grant (Eds.), *Evidence-based Coaching Handbook: Putting Best Practices to Work for Your Clients*, (pp. 313–54). Hoboken, NJ: Wiley.

CBS Studios. (2015). *Star Trek: The Original Series*. Retrieved from http://www.startrek.com/page/star-trek-the-original-series

Centers for Disease Control and Prevention. (n.d.). Attention-deficit/hyperactivity Disorder. Retrieved October 30, 2015 from http://www.cdc.gov/ncbddd/adhd/data.html

Chan, Joseph. (2014). *Confucian Perfectionism: A Political Philosophy for Modern Times*. Princeton, NJ: Princeton University Press

Chandler, Raymond. (1992). *Farewell, my Lovely*. New York, NY: Random House-Vintage. (Original work published in 1940.)

Chaucer, Gregory. (1958). *Canterbury Tales* (Neville Coghill, Trans.). London, England: Penguin Books.

Chomsky, Noam. (2016). *What Kind of Creatures Are We?* New York: Columbia University Press.

Christie, Agatha. (2011). *Murder on the Orient Express: A Hercule Poirot Mystery*. New York, NY: Harper. (Original work published in 1934.)

Cicero. (1971). *On the Good Life* (Michael Grant, Trans.). Harmondsworth, England: Penguin Books.

Collins, Jim. (2001). *Good to Great: Why Some Companies Make the Leap . . . and Others Don't*. New York, NY: HarperBusiness.

Confucius. (n.d.). *The Analects of Confucius* (James Legge, Trans.). Beijing, China: Foreign Language Press.

Connors, C. Keith. (2006). *Attention Deficit Hyperactivity Disorder*. Kansas City, MO: Compact Clinicals.

Covey, Stephen. (2004). *The 7 Habits of Highly Effective People: Powerful Lessons in Personal Change*. New York, NY: Free Press.

Cozby, Paul C. (1997). *Methods in Behavioral Research* (6th ed.). Mountain View, CA: Mayfield.

Csikszentmihalyi, Mihaly. (1990). *Flow: The Psychology of Optimal Experience*. New York, NY: Harper Perennial.

Curd, Patricia (Ed.). (1996). *A Presocratics Reader: Selected Fragments and Testimonia* (Richard D. McKirahan, Jr., Trans.). Indianapolis, IN: Hackett.

Damasio, Antonio. (1994). *Descartes' Error: Emotion, Reason, and the Human Brain*. New York: G. P. Putnam's Sons-Grosset/Putnam.

Daniels, Les. (2000). *Wonder Woman: The Complete History*. San Francisco, CA: Chronicle Books.

Davanloo, Habib. (1992). A method of short-term dynamic psychotherapy. In Habib Davanloo (Ed.) *Short Term Dynamic Psychotherapy*, (pp. 43–71). Northvale, NJ: Jason Aronson.

Davidson, R. J. (2000). Affective style, psychopathology, and resilience: Brain mechanisms and plasticity. *American Psychologist, 55*, 1196–214.

Davidson, S. (1989). *The Effect of Framing on Decision Making in Escalation Scenarios* (Unpublished doctoral dissertation). California School of Professional Psychology/San Diego.

Davidson, S. (2000). An operations model of psychosocial structure and function and of psychotherapy. *Genetic, Social, General Psychology Monographs, 126*(4), 401–57.

Davidson, S. (2011, August). *Is the Mind Subject to Causality? A chaos-based answer*. PowerPoint presented at the international conference the Society for Chaos Theory in Psychology & Life Sciences, Orange, CA.

Davis, Martha, Eshelman, Elizabeth Robbins, & McKay, Matthew. (1982). *Relaxation & Stress Reduction Workbook*. Oakland, CA: New Harbinger.

Dawson, Peg, & Guare, Richard. (2009). *Smart but Scattered: The Revolutionary "Executive Skills" Approach to Helping Kids Reach their Potential*. New York, NY: Guilford Press.

Day, Barry (Ed.). (2014). *The World of Raymond Chandler: In his Own Words*. New York, NY: Knopf.

De George, Richard T., & De George, Fernande. (1972). *The Structuralists: From Marx to Levi Strauss*. Garden City, NY: Doubleday-Anchor.

Dehaene, Stanislas. (2014). *Consciousness and the Brain: Deciphering How the Brain Codes Our Thoughts*. New York: Penguin Random House-Viking.

de Laurentiis, Dino (Producer) & Vidor, King (Director). (1956). *War and Peace* [Motion Picture]. USA: Paramount.

Derrida, Jacques. (1978). *Writing and Difference* (Alan Bass, Trans.). Chicago, IL: University of Chicago Press.

Dewey, John. (1998). *The Essential Dewey: Pragmatism, Education, Democracy* (Vol. 1), (Larry A. Hickman & Thomas M. Alexander, Eds.). Bloomington: Indiana University Press.

Dhammapada: The Path of Perfection. (1973). (Juan Mascaro, Trans.). Harmondsworth, England: Penguin Books.

Dinsmore, Paul C. (1990). *Human Factors in Project Management* (Rev. ed.). New York, NY: AMACOM.

Dollard, John, & Miller, Neal. (1950). *Personality and Psychotherapy: An Analysis in Terms of Learning, Thinking, and Culture*. New York, NY: McGraw-Hill.

Donovan, Dennis M. (1999). Assessment strategies and measures in addictive behaviors. In Barbara S. McCrady & Elizabeth E. Epstein (Eds.), *Addictions: A*

Comprehensive Guidebook, (pp. 187–215). New York, NY: Oxford University Press.

Douglas, Michael (Producer), & Zemeckis, Robert (Director). (1984). *Romancing the Stone* [Motion Picture]. USA: Twentieth Century Fox.

Dumas, Alexandre. (2011). *Three Musketeers*. New York, NY: Knopf-Borzoi. (Original work published in 1844.)

Dwivedi, Kedar Nath (Ed.). (2000). *Post-traumatic Stress Disorder in Children and Adolescents*. London, England: Whurr.

Dylan, Bob. (1964). Mr. Tambourine Man. Retrieved October 23, 2015 from https://www.youtube.com/watch?v=OeP4FFr88SQ

Eliot, T. S. (1930). *The Waste Land and Other Poems.* San Diego, CA: Harcourt Brace Jovanovich.

Ellis, Albert, & Harper, Robert. (1975). *A Guide to Rational Living*. Chatsworth, CA: Wilshire.

Ellis, William. (2004). *A Narrative of an 1823 Tour Through Hawai'i*. Honolulu, HI: Mutual. (Original work published in 1825.)

Emerson, Ralph Waldo. (n.d.). *The Works of Ralph Waldo Emerson* (Vol. 2). New York, NY: Caxton.

Epictetus, & Lebell, Sharon. (1995). *The Art of Living: The Classic Manual on Virtue, Happiness, and Effectiveness*. New York, NY: HarperOne.

Epicurus. (2012). *The Art of Happiness* (George Strodach, Trans.). New York, NY: Penguin Books.

Erasmus. (1971). *Praise of Folly and letter to Martin Dorp 1515* (Betty Radice, Trans.). Harmondsworth, England: Penguin Books.

Ericson, Jon M., & Murphy, James J. (with Zeuschner, Raymond Bud). (1987). *Debater's Guide* (Rev. ed.). Carbondale: Southern Illinois University Press.

Erikson, Erik. (1979). *Childhood and Society* (2nd ed.). New York, NY: Norton.

Euripides. (1981). *Helen* (James Michie & Colin Leach, Trans.). New York, NY: Oxford University Press.

Fawcett, Jacqueline, & Downs, Florence. (1986). *The Relationship of Theory and Research.* Norwalk, CT: Appleton-Century-Crofts.

Ferrer, Sean Hepburn. (2003). *Audrey Hepburn: An Elegant Spirit.* New York, NY: Simon & Schuster-Atria Books.

Feynman, Richard P. (with Leighton, Ralph). (1985). *"Surely You're Joking, Mr. Feynman!": Adventures of a Curious Character.* New York, NY: Norton.

Firkin, Rex (Producer). (1971). *Upstairs, Downstairs* [Television series]. London, UK: London Weekend Television. Retrieved from http://www.imdb.com/title/tt0066722/

Fisch, Richard, Weakland, John, & Segal, Lynn. (1982). *The Tactics of Change: Doing Therapy Briefly.* San Francisco, CA: Jossey-Bass.

Fisher, Roger, & Ury, William. (1981). *Getting to Yes: Negotiating Agreement Without Giving in* (Bruce Patton, Ed.). Harmondsworth, England: Penguin Books.

Fitton, J. Lesley. (2002). *Minoans*. London, England: British Museum Press.

Frank, Jerome D. (1974). *Persuasion and Healing: A Comparative Study of Psychotherapy.* New York, NY: Schocken Books.

Fredrickson, Barbara L. (2009). *Positivity: Top-notch Research Reveals the 3-to-1 Ratio That Will Change Your Life*. New York, NY: MJF Books.

Freedman, Lawrence. (2013). *Strategy: A History*. New York, NY: Oxford University Press.

Freud, Sigmund. (1961). *Civilization and its Discontents* (James Strachey, Trans.). New York, NY: Norton. (Original work published in 1930.)

Freud, Sigmund. (1997). *Dora: An Analysis of a Case of Hysteria*. New York, NY: Simon & Schuster-Touchstone. (Original work published in 1905.)

Fripp, Patricia. (2011). *Presentation Skills Hollywood Style* [DVD]. San Francisco, CA: Chesney Communication.

Galilei, Galileo. (1957). *Discoveries and Opinions of Galileo* (Stillman Drake, Trans.). Garden City, NY: Doubleday-Anchor Books. (Originally published in 1613.)

Gardner, Howard. (1993). *Multiple Intelligences: The Theory in Practice*. New York, NY: Basic Books.

Gardner, Howard. (2011). *Truth, Beauty, and Goodness Reframed: Educating for the Virtues of the Twenty-first Century*. New York, NY: Basic Books.

Garnham, Alan, & Oakhill, Jane. (1994). *Thinking and Reasoning*. Oxford, England: Blackwell.

Gauguin, Paul. (1994). *Noa noa: The Tahiti Journal of Paul Gauguin* (O. F. Theis, Trans.). San Francisco, CA: Chronicle Books.

Geary, D. C. (1995). Reflections of evolution and culture in children's cognition: Implications for mathematical development and instruction. *American Psychologist, 50*, 24–37.

Gebelein, Susan H., Nelson-Neuhaus, Kristie J., Skube, Carol J., Lee, David G., Stevens, Lisa A., Hellervik, Lowell W., & Davis, Brian L. (2004). *Successful Manager's Handbook: Develop Yourself Coach Others* (7th ed.). Minneapolis, MN: Personnel Decisions.

Geller, Bruce (Producer). (1966). *Mission Impossible* [Television series]. Hollywood, CA: Desilu. Retrieved from https://www.youtube.com/watch?v=MA2KmJMKFrQ

Gilligan, Carol. (1982). *In a Different Voice: Psychological Theory and Women's Development*. Cambridge, MA: Harvard University Press.

Gilliland, Burl. (1984). *Theories and Strategies in Counseling and Psychotherapy*. Englewood Cliffs, NJ: Prentice-Hall.

Ginsberg, Allen. (1984). Howl, parts I & II. Poets.org. Retrieved at: https://www.poets.org/poetsorg/poem/howl-parts-i-ii

Gleick, James. (2003). *Isaac Newton*. New York, NY: Random House-Vintage Books.

Goffman, Erving. (1974). *Frame Analysis: An Essay on the Organization of Experience*. New York, NY: Harper-Colophon.

Goffman, Erving. (1986). *Stigma: Notes on the Management of Spoiled Identity*. New York, NY: Simon & Schuster-Touchstone.

Golden, Charles. (1980). *The Luria-Nebraska Neuropsychological Battery: Manual*. Los Angeles, CA: Western Psychological Services.

Goldin, Judah. (1957). *The Living Talmud: The Wisdom of the Fathers*. New York, NY: New American Library-Mentor.

Goodstein, Judith. (1991). *Millikan's School: A History of the California Institute of Technology*. New York, NY: Norton.

Greenberger, Dennis, & Padesky, Christine. (2016). *Mind Over Mood* (2nd ed.). New York, NY: Guilford Press.

Greenhut, Robert (Producer), & Allen, Woody (Director). (1977). *Annie Hall* [Motion Picture]. USA: Metro-Goldwyn-Meyer.

Grodzki, Lynn, & Allen, Wendy. (2005). *The Business and Practice of Coaching: Finding Your Niche, Making Money, & Attracting Ideal Clients*. New York, NY: Norton.

Haidt, Jonathan. (2006). *The Happiness Hypothesis: Finding Modern Truth in Ancient Wisdom*. New York NY: Basic Books.

Hamilton, Nigel. (1992). *JFK: Reckless Youth*. New York, NY: Random House.

Hammer, Michael (with Stanton, Steven A.). (1995). *The Reengineering Revolution: A Handbook*. New York, NY: HarperBusiness.

Hampson, Norman. (1968). *The Enlightenment: An Evaluation of its Assumptions, Attitudes and Values*. London, UK: Penguin.

Hanh, Thich Nhat. (2010). *The Diamond That Cuts Through Illusion*. Berkeley, CA: Parallax Press.

Hanson, Norwood Russell. (1965). *Patterns of Discovery: An Inquiry into the Conceptual Foundations of Science*. Cambridge, England: Cambridge University Press.

Hartmann, Heinz. (1964). *Essays on Ego Psychology: Selected Problems in Psychoanalytic Theory*. New York, NY: International Universities Press.

Hauge, Michael, & Vogler, Christopher. (2003). *The Hero's 2 Journeys* [CD]. Los Angeles, CA: Writer's Audio Shop.

Heatley, C. J., III. (1986). *The Cutting Edge*. Charlottesville, VA: Thomasson Grant & Howell.

Heisenberg, Werner. (1958). *Physics and Philosophy: The Revolution in Modern Science*. New York: Harper & Row-Torchbooks.

Heisenberg, Werner. (1971). *Physics and Beyond: Encounters and Conversations*. New York, NY: Harper-Torchbooks.

Hemingway, Ernest. (1964). *A Moveable Feast* (Restored ed.). New York, NY: Scribner's.

Herman, Arthur. (2001). *The Scottish Enlightenment: The Scots' Invention of the Modern World*. London, England: Harper-Perennial.

Hersen, Michel. (2004). *Psychological Assessment in Clinical Practice: A Pragmatic Guide*. New York, NY: Brunner-Routledge.

Hesiod. (1973). *Theogony* and *Works and Days* (Dorothea Wender, Trans.). London, England: Penguin Books.

Hersey, William D. (1990). *Blueprints for Memory: Your Guide to Remembering Business Facts, Figures, and Faces*. New York, NY: AMACOM.

Hill, Linda. (2003). *Becoming a Manager: How New Managers Master the Challenge of Leadership*. Boston, MA: Harvard Business School Press.

Hirai, Tomio. (1975). *Zen Meditation Therapy*. Tokyo, Japan: Japan.

Homer. (1963). *The Odyssey* (Robert Fitzgerald, Trans.). Garden City, NY: Doubleday-Anchor.

Homer. (2004). *The Iliad* (Robert Fitzgerald, Trans.). New York, NY: Ferrar, Straus & Giroux.

Horney, Karen. (1937). *The Neurotic Personality of our Time*. New York, NY: Norton.

Horney, Karen. (1950). *Neurosis and Human Growth: The Struggle Toward Self-realization*. New York, NY: Norton.

Howard, A. E. Dick. (1998). *Magna Carta: Text and Commentary* (Rev. ed.). Charlottesville: University of Virginia Press.

Hume, David. (1748). *An Enquiry Concerning Human Understanding*. Garden City, NY: Doubleday-Dolphin.

Husserl, Edmund. (1964). *The Idea of Phenomenology*. The Hague: Martinus Nijhoff.

Isaacson, Walter. (2013). *Steve Jobs*. New York, NY: Simon & Schuster.

Ivanov, Iliyan, & Newcorn, Jeffrey. (2005). Attention-deficit/hyperactivity disorders. In Sandra B. Sexson (Ed.) *Child and Adolescent Psychiatry* (2nd ed.), (pp. 91–104). Malden, MA: Blackwell.

Jackendoff, Ray. (2012). *A User's Guide to Thought and Meaning*. Oxford, UK: Oxford University Press.

Jacobson, Torsten Cumberland. (2012). *A History of the Vandals*. Yardley, Pennsylvania: Westholme.

James, William. (1950). *The Principles of Psychology* (Vol. 1). New York, NY: Dover. (Original work published in 1890.)

James, William. (1991). *Pragmatism*. Buffalo, NY: Prometheus Books. (Original work published in 1907.)

Johnson, Joyce. (2012). *The Voice is All: The Lonely Victory of Jack Kerouac*. New York, NY: Viking.

Johansson, Rune E. A. (1969). *The Psychology of Nirvana: A Comparative Study of the Natural Goal of Buddhism and the Aims of Modern Western Psychology*. Garden City, NY: Anchor Books.

Jones, Dawn. (2010). *Taking Control of Time and Priorities: Organizing Your Work and Life* [CD]. Mission, KS: SkillPath.

Jordan, Killian (Ed.). (2000). *Our Finest Hour: The Triumphant Spirit of the World War II Generation*. New York, NY: Time.

Jung, C. G. (1959). *The Archetypes and the Collective Unconscious* (2nd ed., R. F. C. Hull, Trans.). Princeton, NJ: Princeton University Press-Bollingen.

Justinian. (1979). *The Digest of Roman Law: Theft, Rapine, Damage and Insult* (C. F. Colbert, Trans.). London, England: Penguin. (Original work published in 533.)

Kahneman, Daniel. (2011). *Thinking, Fast and Slow*. New York, NY: Ferrar, Straus & Giroux.

Kamauff, John. (2010). *Manager's Guide to Operations Management*. New York, NY: McGraw- Hill.

Kant, Immanuel. (1966). *Critique of Pure Reason* (F. Max Muller, Trans.). Garden City, NY: Anchor Books. (Original work published in 1781.)

Kaufman, Alan, & Lichtenberger, Elizabeth. (2006). *Assessing Adolescent and Adult Intelligence* (3rd ed.). New York, NY: Wiley.

Keene, Carolyn. (1957). *The Hidden Staircase: Nancy Drew Mystery Stories #2*. New York, NY: Grosset & Dunlap.

Kendrick, Tom. (2004). *The Project Management Toolkit: 100 Tips and Technologies for Getting the Job Done Right*. New York, NY: AMACOM.

Kernberg, Paulina F., Weiner, Alan S., & Bardenstein, Karen K. (2000). *Personality Disorders in Children and Adolescents*. New York, NY: Basic Books.

Kerouac, Jack. (1991). *On the Road*. New York, NY: Penguin Books. (Original work published in 1957.)

Kirkland, Stephane. (2013). *Paris Reborn: Napoleon III, Baron Haussmann, and the Quest to Build a Modern City*. New York, NY: St. Martin's Press.

Kolberg, Judith, & Nadeau, Kathleen. (2002). *ADD-friendly Ways to Organize Your Life*. New York, NY: Brunner-Routledge.

Kotz, Samuel, & Stroup, Donna F. (1983). *Educated Guessing: How to Cope in an Uncertain World*. New York, NY: Dekker.

Krauss, Lawrence M. (1993). *Fear of Physics: A Guide for the Perplexed*. New York: HarperCollins-Basic Books.

Kuhn, Thomas S. (1962). *The Structure of Scientific Revolutions*. Chicago, IL: University of Chicago Press.

Laing, R. D. (1967). *The Politics of Experience*. New York, NY: Ballantine Books.

Laing, R. D., & Cooper, D. G. (1964). *Reason and Violence: A Decade of Sartre's Philosophy 1950-1960*. New York, NY: Random House-Pantheon Books.

Lakein, Alan. (1989). *How to Get Control of Your Time and Your Life*. New York, NY: Penguin-Signet.

Landels, J. G. (1978). *Engineering in the Ancient World*. Berkeley: University of California Press.

Lane, Maggie. (1996). *Jane Austen's world: The Life and Times of England's Most Popular Author*. London, England: Carlton Books.

Langley, Noel, Ryerson, Florence, & Woolf, Edgar Allan. (1989). *The Wizard of Oz: The Screenplay* (Michael Patrick Hearn, Ed.). New York, NY: Doubleday-Delta.

Lao Tzu. (1963). *Tao te ching* (D.C. Lau, Trans.). Harmondsworth, England: Penguin Books.

Lax, Eric. (1992). *Woody Allen: A Biography*. New York, NY: Random House-Vintage.

Lear, Norman. (2014). *Even This I Get to Experience*. New York, NY: Penguin Press.

LeDoux, Joseph. (1996). *The Emotional Brain: The Mysterious Underpinnings of Emotional Life*. New York: Simon & Schuster.

Leeming, David Adams. (1990). *The World of Myth: An Anthology*. New York, NY: Oxford University Press.

Leick, Gwendolyn. (2001). *Mesopotamia: The Invention of the City*. London, England: Penguin Books.

Leo, J., & Lacasse, J.R. (2015). The New York Times and the ADHD epidemic. *Society, 52*(1), 3–8.

Lewin, Kurt. (1951). *Field Theory in Social Science: Selected Theoretical Papers*. Cartwright, Dorwin (Ed.). New York: Harper & Row-Torchbooks.

Lezak, Muriel, Howieson, Diane B., & Loring, David W. (with Hannay, H. Julia, & Fischer, Jill S.). (2004). *Neuropsychological Assessment* (4th ed.). Oxford, England: Oxford University Press.

London, Jack. (1989). *Tales of the Pacific*. London, UK: Penguin Books.

Loren, Sophia. (2014). *Yesterday, Today, Tomorrow: My Life* (Sylvia Adrian Notini, Trans.). New York, NY: Simon & Schuster-Atria Books.

Lucas, George (Producer-Director). (1975). *Star Wars: Episode IV – A New Hope* [Motion Picture]. United States: Lucasfilm.

Lucas, George (Producer), & Kershner, Irvin (Director). (1980). *Star Wars: Episode V – The Empire Strikes Back* [Motion Picture]. United States: Lucasfilm.

Luria, A. R. (1973). *The Working Brain: An Introduction to Neuropsychology* (Basil Haigh, Trans.). New York, NY: Basic Books.

Luttwak, Edward N. (1987). *Strategy: The Logic of War and Peace*. Cambridge, MA: Harvard-Belknap.

Lycett, Andrew. (1995). *Ian Fleming*. New York, NY: St. Martin's Press.

Lynch, Peter (with Rothchild, John). (1994). *Beating the Street*. New York, NY: Simon & Schuster.

Macks, Jon. (2003). *How to Be Funny*. New York, NY: Simon & Schuster.

Mager, Robert F. (1997). *Preparing Instructional Objectives: A Critical Tool in the Development of Effective Instruction* (3rd ed.). Atlanta, GA: Center for Effective Performance.

Marean, Curtis W. (2015). The most invasive species of all. *Scientific American, 313*(2), 32–39.

Mariner's Museum. (n.d.). Polynesian navigation. Retrieved from: http://ageofex.marinersmuseum.org/index.php?type=navigationtool&id=10

Martin, Steve. (2001). Changes in the memory after fifty. In David Remnick & Henry Finder (Eds.), *Fierce Pajamas: An Anthology of Humor Writing from the New Yorker* , New York, NY: Modern Library.

Martinson, Jane. (2016, October 25). David Remnick: We are too complicated to just be called serious. *The Guardian*. Retrieved from http://www.theguardian.com/media/2015/oct/25/new-yorker-david-remnick-radio-show

Marx, Karl. (1972). *Capital*. In Richard de George & Fernande de George (Eds.), *Structuralists from Marx to Levi-Strauss,* (pp. 7–23). New York, NY: Doubleday-Anchor.

Maslow, Abraham H. (1968). *Toward a Psychology of Being* (2nd ed.). New York, NY: van Nostrand.

Maslow, Abraham H. (1970). *Motivation and Personality* (2nd ed.). New York, NY: Harper & Row.

Mazzeo, Tilar J. (2012). *Writing Creative Nonfiction*. Chantilly, VA: Great Courses.

McClelland, David. (1967). *The Achieving Society*. New York, NY: McMillan-Free Press.

McClory, Kevin (Producer), & Young, Terence (Director). (1965). *Thunderball.* [Motion Picture]. UK: Eon. Retrieved from https://www.youtube.com/watch?v=R3XSj6kUGkE

McCormack, Mark H. (1986). *What They Don't Teach You at Harvard Business School: Notes from a Street-smart Executive*. New York, NY: Bantam.

McCormick, Ernest. (1979). *Job Analysis: Methods and Applications*. New York, NY: AMACOM.

McInerny, D. Q. (2004). *Being Logical: A Guide to Good Thinking*. New York, NY: Random House.

Meltzoff, Julian, & Kornreich, Melvin. (1970). *Research in Psychotherapy*. New York, NY: Atherton Press.

Merton, Thomas (Ed.). (1960). *Wisdom of the Desert: Sayings of the Desert Fathers of the Fourth Century* (Thomas Merton, Trans.). New York, NY: New Directions.

Miles, Barry. (2013). *Call me Burroughs: A Life*. New York, NY: Hachette-Twelve.

Mill, John Stuart. (1987). *Utilitarianism*. New York, NY: Prometheus Books. (Original work published in 1863).

Miller, George, Galanter, Eugene, & Pribram, Karl. (1960). *Plans and the Structure of Behavior*. New York, NY: Holt.

Miller, William, & Rollnick, Stephen. (2002). *Motivational Interviewing: Preparing People for Change* (2nd ed.). New York, NY: Guilford Press.

Miller, W. R., Zweben, A., DiClemente, C. C., & Rychtarik, R. G. (1992). *Motivational Enhancement Therapy manual: A Clinical Research Guide for Therapists Treating Individuals with Alcohol Abuse and Dependence.* (Vol. 2, Project MATCH Monograph Series) Rockville, MD: National Institute on Alcohol Abuse and Alcoholism.

Milne, A. A. (1926). *Winnie-the-Pooh*. New York, NY: Puffin Books.

Minor, Marianne. (2002). *Coaching and Counseling: A Practical Guide for Managers and Team Leaders*. Menlo Park, CA: Crisp Learning.

Moore, Pete. (2002). *$E=mc^2$: The Great Ideas That Shaped our World.* New York, NY: Metro Books.

Moore, Sonia. (1965). *The Stanislavsky System: The Professional Training of an Actor*. Harmondsworth, England: Penguin Books.

More, Thomas. (1961). *Utopia* (Paul Turner, Trans.). Harmondsworth, England: Penguin Books. (Original work published in 1516.)

Mouskouri, Nana (with Duroy, Lionel). (2007). *Memoirs*. London, UK: Weidenfeld & Nicolson.

Murphy, John (Ed.). (2014). *Showrunners: The Art of Running a TV Show*. [Motion Picture] Dublin, Ireland: Black Sheep Productions.

Musashi, Miyamoto. (1974). *A Book of Five Rings* (Victor Harris, Trans.). Woodstock, NY: Overlook Press.

National Geographic Society. (2010). *Cleopatra: The Search for the Last Queen of Egypt— Official Exhibition Guide*. Nashville, TN: Beckon Books.

National Institute on Alcohol Abuse and Alcoholism. (n.d.). Screening tests. Retrieved October 30, 2015, from http://pubs.niaaa.nih.gov/publications/arh28-2/78-79.htm

Negoita, Constantin Virgil. (1985). *Expert Systems and Fuzzy Systems*. Menlo Park, CA: Benjamin/Cummings.

Neill, A. S. (1972). *Neill! Neill! Orange peel!: An Autobiography by A. S. Neill, the World Famous Headmaster of Summerhill School*. New York, NY: Hart.

Neisser, U. (1976). *Cognition and Reality*. New York, NY: Freeman.

Nelson, Charles, de Haan, Michelle, & Thomas, Kathleen M. (2006). *Neuroscience of Cognitive Development: The Role of Experience and the Developing Brain.* Hoboken, NJ: Wiley.

Nierenberg, Gerard. (1968). *The Art of Negotiating*. New York, NY: Hawthorn Books.

Nietzsche, Friedrich. (1966). *Thus Spoke Zarathustra* (Walter Kaufmann, Trans.). New York, NY: Viking Press.

Nietzsche, Friedrich. (1973). *Beyond Good and Evil: Prelude to a Philosophy of the Future* (R. J. Hollingdale, Trans.). London, England: Penguin. (Originally published in 1886.)

Nisbett, Richard E. (2003). *The Geography of Thought: How Asians and Westerners Think Differently . . . and Why*. New York, NY: Free Press.

Norwood, Robin. (1985). *Women Who Love Too Much: When You Keep Wishing and Hoping He'll Change*. New York, NY: Simon & Schuster-Pocket Books.

Nunnally, Jum C. (1978). *Psychometric Theory*. New York, NY: McGraw-Hill.

O'Flaherty, Wendy Doniger (Ed.). (1981). *The Rig Veda: An Anthology*. London, England: Penguin Books.

Outram, Dorinda. (2006). *Panorama of the Enlightenment*. Los Angeles, CA: Getty.

Parcells, Bill (with Coplon, Jeff). (1995). *Finding a Way to Win: The Principles of Leadership, Teamwork, & Motivation*. New York, NY: Doubleday.

Parente, Rick, & Herrmann, Douglas. (1996). *Retraining Cognition: Techniques and Applications*. Gaithersburg, MD: Aspen.

Pattee, Howard H. (Ed.). (1973). *Hierarchy Theory: The Challenge of Complex Systems*. New York, NY: George Braziller.

Patton, George S., Jr. (1947). *War as I knew it*. New York, NY: Bantam Books.

Paul, Michael. (2000). *Living Zen*. New York, NY: Abbeville Press.

Pavlov, Ivan P. (1960). *Conditioned Reflexes: An Investigation of the Physiological Activity of the Cerebral Cortex*. New York, NY: Dover. (Original work published in 1927.)

Payne, J., Braunstein, M., & Carroll, J. (1978). Exploring predecisional behaviors: An alternative approach to decision research. *Organizational Behavior and Human Performance, 22*, 17–44.

Peirce, Charles. (1955). *Philosophical Writings of Peirce* (Justus Buchler, Ed.). New York, NY: Dover.

Perls, Frederick, Hefferline, Ralph F., & Goodman, Paul. (1951). *Gestalt Therapy: Excitement and Growth in the Human Personality*. New York, NY: Dell-Delta.

Petroski, Henry. (1996). *Invention by Design: How Engineers Get from Thought to Thing*. Cambridge, MA: Harvard University Press.

Phillips, Jack (Ed.). (1997). *Measuring Return on Investment* (Vol. 2). Alexandria, VA: American Society of Training & Development.

Phillips, Larry W. (Ed.). (1984). *Ernest Hemingway on Writing*. New York, NY: Charles Scribner's Sons.

Piaget, Jean, & Inhelder, Barbel. (1969). *The Psychology of the Child* (Helen Weaver, Trans.). New York, NY: Basic Books.

Piskurich, George M. (Ed.). (2002). *HPI: Just-the-facts, Bottom-line Primer on Human Performance Improvement*. Alexandria, VA: American Society for Training & Development.

Plutarch. (1973). *The Age of Alexander* (Ian Scott-Kilvert, Trans.). Harmondsworth, England: Penguin Books.

Polderman, Tinca J. C., Benyamin, Beben, de Leeuw, Christiaan A., Sullivan, Patrick F., van Bochoven, Arjen, Visscher, Peter M., & Posthuma, Danielle. (2015). Meta-analysis of the heritability of human traits based on fifty years of twin studies. *Nature Genetics 47*, 702–09. doi:10.1038/ng.3285

Pollack, Sydney (Producer & Director). (1985). *Out of Africa* [Motion Picture]. USA: Universal.

Polly, Matthew. (2007). *American Shaolin*. New York, NY: Gotham Books.

Polybius. (1979). *The Rise of the Roman Empire* (Ian Scott-Kilvert, Trans.). London, England: Penguin Books.

Porter, Eleanor H. (2003). *Pollyanna*. Mineola, NY: Dover. (Original work published in 1913.)

Posner, Richard. (1990). *The Problems of Jurisprudence*. Cambridge, MA: Harvard University Press.

Post, Peter (with Post, Anna, Post, Lizzie, & Senning, Daniel Post). (2014). *Emily Post's the Etiquette Advantage in Business: Personal Skills for Professional Success* (3rd ed.). New York, NY: HarperCollins-William Morrow.

Powers, William. (1973). *Behavior: The Control of Perception*. Chicago, IL: Aldine.

Puppe, Frank. (1993). *Systematic Introduction to Expert Systems: Knowledge Representations and Problem-solving Methods*. Berlin, Germany: Springer-Verlag.

Pury, Cynthia L. S., & Lopez, Shane J. (Eds.). (2010). *The Psychology of Courage: Modern Research on an Ancient Virtue*. Washington, DC: American Psychological Association.

Quest of the Holy Grail. (1969). (Pauline Matarasso, Trans.). Harmondsworth, England: Penguin Books.

Ragland-Sullivan, Ellie. (1986). *Jacques Lacan and the Philosophy of Psychoanalysis*. Urbana:University of Illinois.

Reynolds, David K. (1976). *Morita Psychotherapy*. Berkeley: University of California Press.

Riggs, Douglas Shepard. (1969). *The Mathematical Approach to Physiological Problems: A Critical Primer*. Cambridge, MA: MIT Press.

Roadstrum, W. H. (1978). *Being Successful as an Engineer*. San Jose, CA: Engineering Press.

Robinson, Andrew. (2010). *Sudden Genius? The Gradual Path to Creative Breakthroughs*. Oxford, UK: Oxford University Press.

Rousseau, Jean-Jacques. (1993). *Emile* (Barbara Foxley, Trans.). London, England: J. M Dent- Everyman. (Original work published in 1762.)

Roszak, Theodore. (1969). *The Making of a Counter Culture: Reflections of the Technocratic Society and its Youthful Opposition*. Garden City, NY: Doubleday-Anchor.

Ruesch, Jurgen, & Bateson, Gregory. (1951). *Communication: The Social Matrix of Psychiatry*.New York, NY: Norton.

Sadock, Benjamin J., & Sadock, Virginia A. (2001). *Kaplan & Sadock's Pocket Handbook of Clinical Psychiatry* (3rd ed.). Philadelphia, PA: Lippincott Williams & Wilkins.

Sagan, Carl. (1980). *Cosmos*. New York, NY: Random House.

Salmond, Anne. (2009). *Aphrodite's Island: The European Discovery of Tahiti*. Berkeley: University of California Press.

Sandler, Mel, & Gray, Muriel. (1999). *Winning at Work: Breaking Free of Personal Traps to Find Success in the New Workplace*. Palo Alto, CA: Davies-Black.

Sartre, Jean-Paul. (1948). *The Emotions: Outline of a Theory*. New York, NY: Citadel Press.

Satir, Virginia. (1967). *Conjoint Family Therapy: A guide to Theory and Technique*. Palo Alto, CA: Science and Behavior Books.

Satow, Ernest. (2006). *A Diplomat in Japan*. Berkeley, CA: Stonebridge. (Original work published in 1921.)

Say, Rosa. (2004). *Managing with Aloha: Bringing Hawaii's Universal Values to the Art of Business*. Honolulu, HI: Ho'ohana.

Schaar, John. (1961). *Escape from Authority: The Perspectives of Erich Fromm*. New York, NY Harper & Row-Torchbooks.

Schafer, Roy. (1983). *The Analytic Attitude*. New York, NY: Basic Books.

Schlenger, Sunny, & Roesch, Roberta. (1989). *How to be Organized in spite of Yourself: Time and Space Management That Works with Your Personal Style*. New York, NY: Penguin-New American Library

Schmidt, F. L., & Hunter, J. E. (1977). Development of a general solution to the problem of validity generalization. *Journal of Applied Psychology, 62*(5), 529–40.

Schoenberg, Richard D. (2004). *The Only Easy Day was Yesterday: Making Navy SEALS*. Annapolis, MD: Naval Institute Press.

Schwartz, Jeffrey, & Begley, Sharon. (2002). *The Mind and the Brain: Neuroplasticity and the Power of Mental Force*. New York, NY: HarperCollins-Regan.

Schwartz, Shalom H., & Bardi, Anat. (2001). Value hierarchies across cultures: Taking a similarities perspective. *Journal of Cross-Cultural Psychology, 32*, 268–90.

Sears, Richard, Rudisill, John, & Mason-Sears, Carrie. (2006). *Consultation Skills for Mental Health Professionals*. New York, NY: Wiley.

Seger, Linda. (2010). *Making a Good Script Great* (3rd ed.). Beverly Hills, CA: Silman-James Press.

Seinfeld, Jerry. (2015, May 31). A laugh is such a pure thing. [Television interview]. New York, NY: CBS. Retrieved from http://www.cbsnews.com/news/jerry-seinfeld-a-laugh-is-such-a-pure-thing/

Selected Collection of the Doctrine of the Mean. (2006). Beijing, China: Sinolingua.

Shakespeare, William. (1993). *Yale Shakespeare* (Wilbur L. Cross, & Tucker Brooke, Eds.). New York, NY: Barnes & Noble Books.

Shaughnessy, Edward L. (Ed.) (2000). *China: Empire and civilization*. Oxford, England: Oxford University Press.

Shelly, Percy Bysshe. (1998). Ozymandias. In Peter Harmon (Ed.), *The Classic Hundred Poems: All-time Favorites* (2nd ed.), (pp. 133). New York, NY: Columbia University Press. (Originally published in 1818.)

Shostrom, Everett. (1968). *Man, the Manipulator: The Inner Journey from Manipulation to Actualization*. New York, NY: Bantam Books.

Silverman, Hugh J., & Ihde, Don. (1985). *Hermeneutics & Deconstruction*. Albany: State University of New York Press.

Simon, Herbert (1976). *Administrative Behavior: A Study of Decision-making Processes in Administrative Organization* (3rd ed.). New York, NY: Free Press.

Simon, Sidney, Howe, Leland W., & Kirschenbaum, Howard. (1972). *Values Clarification: A Handbook of Practical Strategies for Teachers and Students* (Rev. ed.). New York, NY: Hart.

Simonton, Dean Keith. (1994). *Greatness: Who Makes History and Why*. New York, NY: Guilford.

Sinclair, Andrew. (1999). *A Life of Dylan Thomas*. New York, NY: St. Martin's Press-Thomas Dunne Books

Singh, Jagjit. (1966). *Great Ideas in Information Theory, Language and Cybernetics*. New York, NY: Dover.

Sinha, Phulgenda. (1986). *The Gita as it Was: Rediscovering the Original Bhagavadgita*. LaSalle, IL: Open Court.

Skinner, B. F. (1953). *Science and Human Behavior*. New York, NY: Free Press.

Skinner, Quentin (Ed.). (1985). *The Return of Grand Theory in the Human Sciences*. Cambridge, England: Cambridge University Press.

Smart, Bradford. (1999). *Topgrading: How Leading Companies Win by Hiring, Coaching, and Keeping the Best People*. New York, NY: Penguin Books-Portfolio.

Smith, Adam. (1986). *The Essential Adam Smith* (Robert Heilbroner, Ed.). New York, NY: Norton.

Smith, Daniel. (2011). *The Sherlock Holmes Companion: An Elementary Guide*. New York, NY: Castle Books.

Smith, Ian. (2010). *Happy: Simple Steps for Getting the Life You Want*. New York, NY: St. Martin's Griffin.

Spivey, Nigel, & Squire, Michael. (2004). *Panorama of the classical world*. Los Angeles, CA: Getty.

Spock, Benjamin. (1957). *Baby and Child Care* (2nd ed.). New York, NY: Pocket Books.

Springsteen, Bruce. (1978). Darkness on the edge of town. On *Darkness on the edge of town* (track 10). [CD]. New York, NY: Columbia. Retrieved from http://www.azlyrics.com/lyrics/brucespringsteen/darknessontheedgeoftown.html

Stanovich, Keith E., & West, Richard F. (2000). Individual differences in reasoning: Implications for the rationality debate? *Behavioral and Brain Sciences*, *23*, 645–726.

Steinmetz, Katie. (2015, October 26). Help! My parents are millennials. *Time, 186,* 36–43.

Sternberg, Robert J. (2012). Triarchic theory of successful intelligence. In Dawn P. Flanagan & Patti L. Harrison (Eds.), *Contemporary Intellectual Assessment: Theories, Tests, and Issues* (3rd ed.). New York, NY: Guilford Press.

Stratemeyer, Edward. (2007). *The Rover Boys on Treasure Isle*. Fairfield, IA: 1st World Library.

Strozzi-Heckler, Richard. (2007). *The Leadership Dojo: Build Your Foundation as an Exemplary Leader*. Berkeley, CA: North Atlantic-Frog Books.

Strunk, William, & White, E. B. (1979). *Elements of Style* (3rd ed.). New York, NY: Macmillan.

Sullivan, Harry Stack. (1953). *Interpersonal Theory of Psychiatry* (Helen Swick Perry, & Mary Ladd Gawel, Eds.). New York, NY: Norton.

Suzuki, Shunryu. (1970). *Zen mind, beginner's mind* (Trudy Dixon, Ed.). New York, NY: Weatherhill.

Suzuki, Masato, & Katayama, Takuya. (1991, October). *Meta-operations in the Process Model HFSP for the Dynamics and Flexibility of Software Processes.* Washington, DC: IEEE Computer Society. Retrieved from http://scholar.googleusercontent.com/scholar?q=cache:0qKk3Jur4PMJ:scholar.google.com/+suzuki+katayama+1991+meta-operations&hl=en&as_sdt=0,5&as_vis=1

Taylor, Val. (2002). *Stage Writing.* Ramsbury, England: Crowood Press.

Thomas, Dylan. (1954). *Under Milk Wood: A Play for Voices.* New York, NY: New Directions.

Thomas, Dylan. (1998). Do not go gentle into that good night. In Peter Harmon (Ed.), *The Classic Hundred Poems: All-time Favorites* (2nd ed.), (pp. 251). New York, NY: Columbia University Press. (Original work published in 1952.)

Thoreau, Henry David. (2008). *Walden.* New York, NY: Fall River Press.

Thoresen, Carl E., & Mahoney, Michael J. (1974). *Behavioral Self-control.* New York, NY: Holt, Rinehart and Winston.

Tolstoy, Leo. (1993). *War and Peace.* New York, NY: Barnes & Noble.

Tomes, Anne, & Hayes, Mike. (1993). *Operations Management: Principles and Practice.* New York, NY: Prentice Hall.

Toropov, Brandon. (1999). *Manager's Portfolio of Model Performance Evaluations: Ready-to-use Performance Appraisals Covering all Employee Functions.* Paramus, NJ: Prentice Hall.

Trump, Ivanka. (2009). *The Trump Card: Playing to Win in Work and Life.* New York, NY: Simon & Schuster-Touchstone.

Twain, Mark. (1995). *Mark Twain's Book for Bad Boys and Girls* (R. Kent Rasmussen, Ed.). Chicago, IL: Contemporary Books.

Twain, Mark. (2006). *The Prince and the Pauper.* Retrieved from http://www.gutenberg.org/files/1837/old/orig1837-h/main.htm

Twelve Steps and Twelve Traditions. (1953). New York, NY: Alcoholics Anonymous World Services.

Upanishads: Translations from the Sanskrit. (1965). (Juan Mascaro, Trans.). London, England: Penguin Books.

Upton, Albert. (1941). *Design for Thinking: A First Book in Semantics.* Stanford, CA: Stanford University Press.

Vaillant, George E. (1977). *Adaptation to Life.* Boston, MA: Little, Brown.

van Genderen, Hannie, Rijkeboer, Marleen, & Arntz, Arnoud. (2012). Theoretical model: Schemas, coping styles, and modes. In *The Wiley-Blackwell Handbook of Schema Therapy: Theory, Research, and Practice* (Michiel van Vreeswijk, Jenny Broersen, & Marjon Nadort, Eds.), (27–40). Malden, MA: Wiley-Blackwell.

von Bertalanffy, Ludwig. (1968). *General System Theory: Foundation, Development, Applications.* New York, NY: George Braziller.

von Neuman, J., & Morgenstern, O. (1953). *Theory of Games and Economic Behavior* (3rd ed.). Princeton, NJ: Princeton University Press.

Waldrop, M. Mitchell. (1992). *Complexity: The Emerging Science at the Edge of Order and Chaos.* New York, NY: Simon & Schuster.

Walker, Stuart. (1986). *Winning: The Psychology of Competition.* New York, NY: Norton.

Wallerstein, Judith S., & Blakeslee, Sandra. (1995). *The Good Marriage: How and Why Love Lasts.* Boston, MA: Houghton Mifflin.

Wallis, Hal (Producer), & Curtiz, Michael (Director). (1942). *Casablanca.* [Motion picture]. USA: Warner Brothers.

Warren, James. (2007). *Presocratics: Natural Philosophers Before Socrates.* Berkeley: University of California Press.

Warring, R. H. (1984). *Logic Made Easy.* Guildford, England: Lutterworth Press.

Watson, John B. (1970). *Behaviorism.* New York, NY: Norton. (Original work published in 1924.)

Watzlawick, Paul, Bavelas, Janet Beavin, & Jackson, Don D. (1967). *Pragmatics of Human Communication: A Study of Interactional Patterns, Pathologies, and Paradoxes.* New York, NY: Norton.

Watzlawick, Paul, Weakland, John, & Fisch, Richard. (1974). *Change: Principles of Problem Formation and Problem Resolution.* New York, NY: Norton.

Weatherford, Roy. (1982). *Philosophical Foundations of Probability Theory.* London, England: Routledge & Kegan Paul.

Wechsler, David. (2008). *Wechsler Adult Intelligence Scale* (4th ed.). San Antonio, TX: Pearson.

Weinberg, Gerald. (1985). *The Secrets of Consulting: A Guide to Giving & Getting Advice Successfully.* New York: Dorset.

Weinberg, Victor. (1980). *Structured Analysis.* Englewood Cliffs, NJ: Prentice-Hall.

Wheelis, Allen. (1966). *The Illusionless Man.* New York, NY: Norton.

White, Pepper. (1991). *The Idea Factory: Learning to Think at MIT.* New York, NY: Penguin-Plume.

White, Robert. (1959). Motivation reconsidered: The concept of competence. *Psychological Review, 66,* 297–333.

Whitmore, John. (1992). *Coaching for Performance: Growing People, Performance and Purpose* (3rd ed.). London, England: Nicholas Brealey.

Wiener, Norbert. (1961). *Cybernetics: Or Control and Communication in the Animal and the Machine* (2nd ed.). Cambridge, MA: MIT Press.

Wilde, Oscar. (1986). *Oscar Wilde: The Importance of Being Earnest and Other Plays.* London, England: Penguin Books.

Wilder, Lilyan. (1986). *Professionally Speaking: Getting Ahead in Business and Life Through Effective Communicating.* New York, NY: Simon and Schuster.

Willerding, Margaret F., & Hayward, Ruth A. (1968). *Mathematics: The Alphabet of Science.* New York, NY: Wiley.

Willner, Ann Ruth. (1984). *Spellbinders: Charismatic Political Leadership.* New Haven, CT: Yale University Press.

Wills, Chuck. (2009). *Jack Kennedy: The Illustrated Life of a President.* San Francisco, CA: Chronicle Books.

Wilmshurst, Linda. (2005). *Essentials of Child Psychopathology.* Hoboken, NJ: Wiley.

Wilson, Arnold, & Gedo, John E. (1993). *Hierarchical Concepts in Psychoanalysis: Theory, Research, and Clinical Practice.* New York, NY: Guilford Press.

Wilson, R. Andrew. (2009). *Write Like Hemingway.* Avon, MA: Adams.

Wister, Owen. (1979). *The Virginian: A Horseman of the Plains.* New York, NY: New American- Signet. (Original work published in 1902.)

Wokler, Robert. (1995). *Rousseau*. Oxford, England: Oxford University Press.
Wolin, Steven, & Wolin, Sybil. (1993). *The Resilient Self: How Survivors of Troubled Families Rise Above Adversity*. New York, NY: Random House-Villard.
World Health Organization. (1992). *The ICD-10 Classification of Mental and Behavioural Disorders: Clinical Descriptions and Diagnostic Guidelines*. Geneva, Switzerland: Author.
Wright, George. (1984). *Behavioral Decision Theory: An Introduction*. Beverly Hills, CA: Sage.
Wyss, Johann David. (2006). *Swiss Family Robinson*. New York, NY: Sterling. (Original work published in 1812.)
Yamamoto, Tsunetomo. (2008). *The Art of the Samurai: Yamamoto Tsunetomo's Hagakure* (Barry Steben, Trans.). London, England: Duncan Baird Publishers.
Yapko, Michael. (1990). *Trancework: An Introduction to the Practice of Clinical Hypnosis* (2nd ed.). Florence, KY: Brunner/Mazel.
Yourdon, Edward, & Constantine, Larry L. (1979). *Structured Design: Fundamentals of a Discipline of Computer Program and Systems Design*. Englewood Cliffs, NJ: Prentice Hall- Yourdon Press.
Yuzan, Daidoji. (1988). *The Code of the Samurai: Budo shoshinshu* (A. L. Sadler, Trans.). Rutland, VT: Charles E. Tuttle. (Original work published in 1941.)
Zeeman, E. C. (1976). Catastrophe theory. *Scientific American*, *234*(4), 65–83.
Zegarelli, Mark. (2007). *Logic for Dummies*. Hoboken, NJ: Wiley.
Zi, Nancy. (1994). *The Art of Breathing: A Course of Six Simple Lessons to Improve Performance and Well-being*. Glendale, CA: Vivi.

Index

About the Author

Steve Davidson, PhD, is a licensed clinical psychologist with a specialization in systems theory and learning within an operations frame. He received his PhD from the California School of Professional Psychology, San Diego. He has experience in staff training, teaching graduate students, and government consulting, and maintains an independent practice. As a systems theorist, his research interests include learning from those past cultures, which evidence high levels of functioning and which might serve as models for contemporary societies.

But all of this he considers his day job, intended to support his true identity as a philosopher in a café in Paris.